T0155836

Statistical and Multivariate Analysis in Material Science

Editor

Giorgio Luciano

National Research Council of Italy
Institute for the Study of Macromolecules
Genova, Italy

CRC Press is an imprint of the
Taylor & Francis Group, an **Informa** business
A SCIENCE PUBLISHERS BOOK

Cover credit: Images drawn by Giorgio Luciano. Part of the images reproduced on the cover are based on the dataset Datasaurus by Justin Matejka (Sr. Principal Research Scientist, Toronto, Canada), George Fitzmaurice (Director, Human Computer Interaction and Visualization Research Group, Toronto, Canada).

First edition published 2021
by CRC Press
6000 Broken Sound Parkway NW, Suite 300, Boca Raton, FL 33487-2742

and by CRC Press
2 Park Square, Milton Park, Abingdon, Oxon, OX14 4RN

CRC Press is an imprint of Taylor & Francis Group, LLC

Library of Congress Cataloging-in-Publication Data

Names: Luciano, Giorgio, 1975- editor.
Title: Statistical and multivariate analysis in material science / Giorgio Luciano, National Research Council of Italy, Institute for the Study of Macromolecules, Genova, Italy.
Description: First edition. | Boca Raton : CRC Press, Taylor & Francis Group, 2021. | Includes bibliographical references and index.
Identifiers: LCCN 2020043491 | ISBN 9781138196308 (hardcover)
Subjects: LCSH: Materials science--Statistical methods. | Multivariate analysis.
Classification: LCC TA404.3 .S73 2021 | DDC 620.1/1015195--dc23
LC record available at https://lccn.loc.gov/2020043491

ISBN: 978-1-138-19630-8 (hbk)

Typeset in Times New Roman
by Radiant Productions

Preface

WHO IS THE BOOK ADDRESSED TO

The present work is an introductory text in statistics, addressed to researchers and students in the field of material science. It has the aim of giving the readers a basic knowledge on how statistical reasoning is exploitable in this field, improving their knowledge of statistical tool, and helping them to carry out statistical analyses and to interpret the results.

It also focuses on establishing a consistent multivariate workflow starting from a correct design of experiment followed by a multivariate analysis process.

STYLE OF THE BOOK

Every student/reader with a basic knowledge of algebra and calculus could read the book without problems since it is suitable for graduate and undergraduate students of courses of Chemistry, Physics, Material Science, and every course that deals with synthesis and characterization of materials. The style of the book is informal and tries to use only minimal mathematical notation, avoiding detailed description of the theoretical background of the statistical issues presented. Readers will be introduced to a range of sample cases picked up from real research work in the field of material science in order to stimulate their interest in applying the learned topics and train via exercises. The book indeed has the purpose of explaining the matter of the statistical reasoning over the mathematical calculations and formulations in order to make the readers aware of the importance of the statistical design in their experiments, the type of the data distributions, the meaning of statistical significance of the results, the usefulness of clustering and regression techniques.

USE OF COMPUTER

R statistical language, due to its open-source/free nature and its vast adoption in the field of scientific disciplines, is the language of choice of this book. Nevertheless, the approach to separate the code written to analyze the data from the main text was chosen in order let readers/students have the freedom to use their own favorite software to implement in their workflow.

It is important to remember that the availability of the remarkable computing power makes it more important that the researcher applies statistical methods rationally and correctly: the use of the computer isn't advantageous if it isn't intelligent. To be familiar with the meaning of the statistics is essential to be able to give appropriate input and to interpret the output accurately.

Readers are invited to read the software section, even if they are not versed in using a scripting language/programming, in order to understand the way of thinking on the basis of a scripted workflow. All the scripts and data sets can be downloaded from the author repository at the address https://github.com/jojosgithub/ESMAM. The author will be happy to hear feedback and exchange ideas if contated via email at giorgio.luciano@cnr.it

ORGANIZATION

The idea of the book is to encourage the reader to build a solid and consistent statistical workflow in the development, synthesis, and characterization of materials. Each chapter is structured in the following way:

1. List of the objective focus of the chapter
2. Introduction
3. Description of the data-set(s) used in the examples
4. Description and application of the methodologies
5. Case study(ies)
6. Frequently Asked Questions
7. Remarks (including resources for the students and template for exercise that the student should perform in order to train on the methodologies presented)
8. Bibliography

The subjects of the chapters of the book are the following:

1. Description and validation of the acquired data by means of descriptive and inferential statistics focusing on Analysis of variance (ANOVA) and inferential tests and the visualization of the data gathered using exploratory data technique
2. Application of Design of experiment (DOE) in order to reduce the number of experiments to perform and maximize the information on the system under study
3. Explorative data analysis by means of multivariate techniques with a particular attention to clustering techniques
4. Modelling and regression techniques by means of Partial least squares (PLS) and similar tools
5. Appendix 1: Statistical tables
6. Appendix 2: Design of experiment tables
7. Appendix 3: Commented code used to generate the examples for each chapter

All the chapters are independent from each other in order to let the reader focus on the most relevant aspect of his own research.

Real life commented case studies will let the reader get acquainted with the new tools presented, and the reader is suggested to try to solve the exercises in each chapter and compare his own solutions with those reported.

WHY A BOOK ABOUT MATERIAL SCIENCE AND STATISTICS

The idea of writing this book started while I was a student attending data analysis and statistics courses at the department of Chemistry at the University of Genoa. I enjoyed the courses, but I inevitably felt a kind of divide between the exercises shown and the work in laboratory.

The majority of exercises were mainly about car manufacturers, agriculture, food science, pharmaceutical drug design, social sciences. The exercises closest to the laboratory practice came from analytical chemistry (i.e., meaning and standard deviations of the molarity of a batch of samples, comparing batches, finding outliers from batches, etc.).

After graduating and during my PhD, due to the multivariate nature of the analysis that I performed, I started learning basic concepts of Chemometrics, thanks to Prof. Riccardo Leardi, who organized a workshop for PhD students on design of experiment and multivariate analysis. The data-set was mainly focused on food and pharmaceuticals during the course, but indeed it was a big step ahead in comparison with the previous courses, and there was a common ground about spectroscopy due to my necessity of analyzing results via multivariate analysis.

I then got more involved in data analysis, and after my PhD, I was lucky enough to spend a year as a postDoc at Matforsk (now Nofima at Aes, Oslo) with Prof. Tormoed Naes. It was very stimulating and it gave me the chance to better understand the different approach statisticians and chemists have in the field of multivariate analysis and how a combined approach is always necessary in an interdisciplinary environment. At that time the common ground was sensorimetrics and again multivariate statistics.

After a while, I came back to chemistry, and more specifically, to the study of metal corrosion, trying to apply the statistical techniques I learned, and I found a niche in the simple applications of cluster analysis. Eventually, the research expanded also to coatings, polymers, and composite materials.

My focus became: "*multivariate analysis applied to material science*". After a fortuitous case I met the publisher of this book (who of course I would like to thank), and after talking about a few ideas, I had the green light to write a book that I would have loved to read when I was a student. A book of statistics for scientist with examples from their own field of study. And so with the help of the co-authors of this book, I started to organize my experience (and their contributes) in the form of the book you are reading.

I admit that at the beginning my first problem was to define the field of *Material Science*. I had not questioned the idea that I was working in material science before. As always I started from literature to get help.

Having a look at the material science section in a mainstream publisher website presented (and will still present) journals and books ranging in subject areas that include Engineering, Physics, Medicine, Chemistry (Organic, Inorganic, Analytical and Physical), Chemical Engineering, Environmental Science, Energy. My idea got even more confused, and so I failed to write a generic and inclusive definition of material science, and it inevitably ended up being too narrow or too generic. My interested in finding a way to define material science increased and so I started to have a look at the *history of material science* to find the roots of this discipline.

It is not easy to trace the origin of material science and establish when it became a recognized discipline. Almost all researchers in this field agree on the fact that it developed from metallurgy in the 50s.

R.W. Cahn, in his book "The Coming of Materials Science", describes the emergence of this new discipline, starting from metallurgy, tracing its first steps in the departments of the University of California, Northwestern University, Massachusetts Institute of Technology, University of Wales, and University of Sussex, and its following development in the research laboratories of Dupont, Bell, and General Electric, where the field of research expanded in order to include ceramics and polymers.

What was the common aim of all those departments and laboratories? The pursue of the study was to investigate the structure of matter at different dimensions (from macroscopic to atomic) and relate it to its physical and chemical properties. Nowadays, after more than half a century, this is still the core of material science and concerns the interdisciplinary study of materials.

We characterize natural materials in order to find inspiration for new synthesis in laboratory, investigate the structure of synthesized materials, and study their properties (e.g., mechanical, electrical, thermal, just to name a few) helped by advanced techniques that let us gather data sets that are increasing in dimension by orders of magnitudes.

For historical reasons, laboratories focusing on materials have preferred subject (e.g., metals, ceramics or organic), but the boundaries are always merging, and it is the norm to work in a multidisciplinary environment. About the everyday job of a material scientist it comes from itself that a researcher dealing with the study of materials would need to plan experiments trying to gather as much information as possible by means of several different techniques and merge the results in a thoughtful way in order to get better conclusions.

In this framework, several statistical tools can be helpful to achieve this task.

But how should we define statistics in order to include its contribution to material science? A useful and widespread definition that statistics is the science that collects, analyzes, interprets, and presents data with the pursuit of understanding them, and of making decisions despite variability and uncertainty is valuable from our point of view.

Statistical methods let researchers plan experiments taking variability into account and extract the maximum information from the data and quantify the reliability of information. During the first phase of data gathering, Design of Experiments minimizes the number of the analyses to perform while retaining the maximum information possible and then univariate and multivariate statistics let us take the best decisions with the gathered information.

Attention should be paid since these tools do not substitute the necessary knowledge of Physics, Inorganic and Organic Synthesis, Physical and Analytical Chemistry, that a material scientist needs to have, but are the indispensable complements to a correct workflow.

The blind use of software tool in data analysis is common, and it can amplify the risk of jumping straight to the calculations without looking at the data and asking the right questions about them, forgetting that we have to analyze, understand, and interpret data, and not just perform calculations. The goal of this book is to bring the focus of readers to these fundamental aspects.

Acknowledgements

First of all I would like to thank my editor for his patience and helping me to keep being on tracks. I would like to thank Tormod Naes since he taught me to have long term goals in my scientific path, like writing this book. Richard Brereton for helping me keeping the pace, Riccardo Leardi who was the first to introduce me to Chemometrics. All my collegues at the Institute for Chemical Technologies at the National Research Council of Italy with a special thank to Ilaria Schizzi, all my friends that still talk to me after avoiding to meet them during several weekends. This book is dedicated to my parents, my sister, my grandpas and gradmas for their support and unconditioned love and patience.

Contents

Appendices

Part I

Statistics Basics

1 Statistics Basics

Objectives:

- Discuss the concept of bias and precision
- Refresh basic concepts of statistics
- Apply graphics representation of data for performing exploratory data analysis (EDA)
- Suggest a few caveats while applying basic statistics techniques to material science

1.1 INTRODUCTION

During all the processes of the synthesis and characterization of a material, we will have to face challenges that will depend on their data variability.

A few examples can be:

- Quantify the effect of a reagent
- Study the effect of a process variable
- Evaluate the properties of a newly synthesized material in comparison with an old one

In these processes, we will also need to discern patterns and carefully plan experiments in order to minimize their number, and finally extract as much information as possible from the data analysis performed.

The following data-set will be used to illustrate the challenges for the material scientist related to the variability found in data-sets.

1.1.1 DATA-SET PLA

Data-set is based on a publication by La Mantia et al. (6), where they investigate the effect of the presence of small amounts of PLA on the recycling of PET bottles by several techniques. It consists of 6 samples $\times$ 4 sets of replicates. Each sample is measured 16 times, and the variables studied are four. We will describe the meanings of all these terms later in the chapter.

1.1.2 DATA-SET FGO

Data-set based on a work by Hu et al. (3). The authors evaluated the effect on the scratch resistance and other mechanical properties of incorporating Functionalized

Table 1.1
Summary of data-set PLA based on (6). Effects of the presence of small amounts of PLA on the Elastic Modulus, Tensile Strength, and Elongation at break of recycled PET bottles.

id	Content (wt%)	Elastic Modulus (MPa)	Tensile Strength (MPa)	Elongation at Break (%)
S1	0	1280 ± 60	36 ± 2	450 ± 20
S2	0.5	1240 ± 60	35 ± 2	390 ± 20
S3	1	1400 ± 70	34 ± 2	430 ± 20
S4	2	1350 ± 70	34 ± 2	400 ± 20
S5	5	1400 ± 100	29 ± 2	380 ± 30
S6	100	1250 ± 90	43 ± 2	3.4 ± 0.3

Graphene Oxide (FGO) in comparison with graphene oxide into polysiloxane coatings. It consists of 6 samples × 4 sets of replicates. Each sample is measured 16 times and the variables studied are four.

Table 1.2
Data-set FGO based on (3), reporting the effect on the scratch resistance and other mechanical properties of incorporating Functionalized Graphene Oxide into polysiloxane coatings.

id	Content (wt%)	Pencil Hardness	Flexibility (mm)	Scratch Resistance (g)	Hardness (Mpa)	Elastic Modulus (Gpa)
Filler	0	3H	15	600	167 ± 4	2.20 ± 0.02
GO	0.25	3H	15	600	159 ± 2	2.38 ± 0.08
GO	0.5	3H	15	600	175 ± 2	2.2 ± 0.2
GO	0.75	4H	15	700	212 ± 7	2.2 ± 0.1
GO	1	3H	>15	700	180 ± 10	2.2 ± 0.2
FGO	0.25	4H	15	700	201 ± 6	2.20 ± 0.02
FGO	0.5	4H	15	800	221 ± 5	2.20 ± 0.02
FGO	0.75	5H	15	1100	230 ± 10	2.20 ± 0.02
FGO	1	5H	15	900	240 ± 10	2.20 ± 0.02

1.2 SAMPLES AND VARIABLES

Before focusing on the objectives of this chapter, it is useful to introduce a few statistical definitions. A data-set where *observations* are taken under homogeneous conditions is called a *sample*. All the *samples* that we can take form the *population*.

The number of observations in a sample is called the *sample size*, and is denoted by an *n*. As an example:

- Glass transition temperature of *n* composites
- Young's modulus of *n* epoxy resins
- Tensile strength of *n* novel synthesized polymers
- Compressive modulus of *n* hybrid composites of a rubbery epoxy resin

in these case the *observed variable* are the:

- Glass transition temperature
- Young modulus
- Tensile strength
- Compressive modulus

and the *observational units* are the:

- Composites
- Epoxy resins
- Novel synthesized polymers
- Rubbery epoxy resin

The *sample size* is *n*. The observation that we can perform on each observational unit can be a qualitative observation of a *categorical variable* or a *quantitative observation*.

Qualitative variables describe a quality or characteristic on each experimental unit, while *quantitative variables* measure a numerical quantity or amount in each experimental unit.

Qualitative variables produce categorical data, while quantitative variables produce both discrete and continuous data.

A *discrete variable* can assume only finite values, while a continuous variable can assume infinite values in an interval.

Examples of *qualitative variables* are the:

- *Catalyzer* used for the synthesis of a coating
- *Organic binder* used for manufacturing a refractory ceramic brick
- *Polymeric* blend used for blowing bottles
- *Process* employed for creating a polymer blend

Examples of *quantitative* variable are the:

- *Concentration* of the catalyzer
- *Concentration* of organic binder used for manufacturing a refractory ceramic brick
- *Concentration* of each component of the polymeric blend used for blowing bottles

- *Value of a parameter*, i.e., rpm, temperature, and pressure varied in the process employed for creating a polymer blend

The distinction between continuous and discrete variables is not a rigid one. After all, physical measurements can always be rounded off.

As an example, when in a chemical system, a *temperature*, a *pressure*, a *concentration*, etc. can vary in a predefined numeric range, we have *quantitative continuous variables*, while when they can vary only in discrete intervals, we have *quantitative discrete variables* (i.e., a heater with a control gauge that has only three positions- low, med, and high). The *type* of catalyzer is a categorical value (i.e., catalyzer A and/or B) and its *concentration* can be a *quantitative continuous* of a discrete quantitative variable.

Univariate data are the result of observations performed on a single variable while, when we measure more than two variables, we obtain a *Multivariate data-set*.

Think about the last experiments that you performed in the laboratory

- What were the variables involved?
- Where were qualitative or quantitative?
- What were the subsets of thc population of your data?

As suggested by IUPAC (Union of Pure and Applied Chemistry), the term *sample* is defined as:

A portion of material selected from a larger quantity (of material).

Whole *specimen* is defined in analytical chemistry as:

A specifically selected portion of a material taken from a dynamic system and assumed to be representative of the parent material at the time it is taken.

As an example:

- Glass transition temperature of a *specimen* from a *sample* of composites materials
- Young's modulus of a *specimen* from *sample* epoxy resins
- Tensile strength of a *specimen* from a *sample* of novel synthesized polymers
- Compressive modulus of a rubbery epoxy resin *specimen* from a *sample* of hybrid composites

After the introduction of the previous definition we can write that:

- Data-set PLA records the measurement performed on six *samples*. All the *variables* are presumed to be *continuous* and each *sample* was subdivided to five *specimens*.
- Data-set FGO presents three groups of *samples* (Filler, GO, and FGO) with different contents of an additive (GO and FGO). Three *continuous variables* were monitored (Elastic Modulus, Hardness, Scratch Resistance, and percentage of weight Content of additive), and two *categorical* Pencil Hardness and Flexibility).

1.3 ERRORS

The first thing to keep in mind is that our results are worthless if we do not give the reader an estimate of their errors. We can define the *absolute error*, E_A, in a measurement or result, x_m as:

$$E_A = x_m - x_t$$

where x_m is the true or accepted value, while the *relative error* is given by:

$$E_R = (x_m - x_t)/x_t$$

E_r is often expressed as relative error $100E_R$ since it is useful when comparing results of different magnitudes.

This is true for *every measurement we perform* that gives us as output a numeric value. It is also true that *every analysis* that we perform involves dealing with the concept of *reproducible research* and its *associated error.*

We can distinguish three types of errors. The first one, *gross error*, can be easily recognized and ruled out. It's such a big error that we need to reconsider and perform our experiment or measurement again (i.e., mislabeled samples, wrong calibration, wrong reagent during the synthesis, and so on). We can split the remaining contributes into *random* and *systematic*.

First of all, we need to check the presence of *random errors* that cause the replicates to differ from one another. The wider the span of the measurements, the less precise we were in performing our analyses. When measurements are higher (or lower) in comparison to the true value, we are in the presence of a systematic error, and the source of the systematic error is the bias of our measurement(s).

Random errors affect *precision* and can be caused by both the experimenter and/or the equipment and can be estimated and minimized.

Systematic errors produce a *bias*, so there is a deviation from the true result even if the random errors are very small.

Measurement replicates cannot help in avoiding them. They affect all the results increasing or decreasing the true value in one sense exclusively, they can be avoided using standardized methodology and materials. The experimenter, the equipment, or the methodology employed in running the experiment/analysis can be the cause of them. *Systematic errors* can be classified as *operator errors*, *method* or *procedural errors*.

We can succeed in converting the method to an *unbiased* and more *precise* one, while *random* errors are unavoidable by their own definition.

The first thing that we can do to avoid systematic error is to know in advance if the methodology that we are employing is prone to error (i.e., sampling in a contaminated environment); a second very common cause of error is wrongly calibrated instruments.

Another cause of systematic error can be the operator itself, i.e., a wrong digit in a spreadsheet copied from file to file or non-standard procedures in rounding digits. *Systematic Errors* can be *proportional* when they depend on the size of the sample measured or *constant* when they are independent of sample size (8; 11).

1.4 INITIAL DATA ANALYSIS

We will now present a general work-flow for Initial Data Analysis by means of a commented case study based on the data presented in Table 1.1 for the data-set PLA. We refer to 4 *sets* of experiments performed on 36 *specimens* of a *sample* to determine the tensile strength.

Table 1.3
Tensile strength value for labs A, B, C, and D reported with raw numbers of digits from the collected measurements.

Sample	Tensile Strength
A	32.17, 38.11, 41.34, 27.62, 38.72, 39.02, 34.7, 34.81, 34.74, 33.44, 35.09, 33.01, 33.89, 37.26, 40.84, 36.56, 34.96, 33.36, 33.65, 46.66, 37.54, 35.04, 35.24, 38.84, 34.23, 31.21, 39.3, 32.91, 36.94, 33.26, 41.41, 35.1, 34.16, 34.99, 30.48, 32.33
B	46.23, 34.61, 40.18, 41.8, 40.43, 37.36, 47.07, 37.43, 50.11, 37.62, 45.83, 51.72, 29.67, 36.33, 37.2, 41.82, 36.29, 22.06, 23.36, 45.92, 36.16, 27.31, 36.97, 45.29, 49.37, 35.42, 36.46, 27.42, 40.76, 34.16, 40.73, 42.23, 44.21, 34.35, 41.03, 27.7
C	38.94, 33.61, 39.07, 33.61, 45.83, 39.73, 37.22, 42.65, 34.76, 33.73, 34.17, 36.32, 24.71, 35.8, 49.87, 41.83, 41.79, 32.74, 40.03, 39, 23.82, 37.37, 31.3, 43.64, 44.63, 38.56, 25.53, 35.85, 27.07, 32.75, 39.97, 41.15, 43.51, 28.71, 45.14, 28.58
D	36.87, 32.75, 41.18, 39.08, 36.89, 35.74, 30.49, 32.93, 34.13, 33.52, 43.3, 38.54, 41.75, 38.69, 41.83, 32.55, 38.27, 38.4, 41.61, 29.7, 33.09, 40.57, 36.56, 39.18, 33.5, 40.57, 33.57, 34.46, 31.78, 37.49, 47.53, 40.41, 38.72, 42.32, 37.01, 29.55

1.4.1 SIGNIFICANT DIGITS

A few considerations about how to report data. How many digits do we need to write down in our table? We need to be sure that each digit is a *significant* one.

What does this mean? Let us say that you measure the tensile strength of a coating, and that it is of 35.5 $\pm$ 0.4 (Mpa). You are unsure of the last digit, but you probably know that your number is nearer to 36 than to 37 or 39. If you would like to report only the *significant digits* you would report only 35.

Zeros are significant when they occur between non-zero digits, and they are not significant when they only locate the decimal point.

How to deal with a measurement with a reported value of 0.002500? There are four significant digits since the first three zeros are there only to locate the decimal point, while the final two are not necessary to locate the decimal point, and so they must have been included because the number is known with sufficient accuracy.

Let us see another example. What are the significant digits if I report that the tensile strength of a nanocomposite is 30 Mpa?

In this case it depends on the knowledge about significant digits of the experimenter that reports this measurement. We can assume that there are only two, and for clarity's sake the result should have been reported as 30. In order to avoid any problems, it would be better indeed to report all the results using *scientific notation*.

In scientific notation every number is reported using *two factors*. The *first one* is a number between 1 and 10 and the *second one* is a power of ten. 30 in the previous example can be written as $3 \cdot 10$ if the 0 was not significant or $3 \cdot 0$ if the 0 digit was significant.

We finally need to introduce the concept of *rounding*. You should report the closest number with the proper significant digits to the calculated value, i.e., 35.2 with two significant digits is rounded to 35 and 35.6 to 36. If the number is 5, you generally round to the even digit (35.5 would be rounded to 36, while 34.5 will rounded to 34).

A practical convention, is to use all the digits you obtain from the instruments and use the correct number of significant digits only at the end of calculations.

Table 1.4
Tensile strength value for labs A, B, C, and D, performed on sample S1 from Table 1.2, reported with a raw number of digits from the collected measurements (Mpa).

Sample	Tensile Strength
A	32, 38, 41, 28, 39, 39, 35, 35, 35, 33, 35, 33, 34, 37, 41, 37, 35, 33, 34, 47, 38, 35, 35, 39, 34, 31, 39, 33, 37, 33, 41, 35, 34, 35, 30, 32
B	46, 35, 40, 42, 40, 37, 47, 37, 50, 38, 46, 52, 30, 36, 37, 42, 36, 22, 23, 46, 36, 27, 37, 45, 49, 35, 36, 27, 41, 34, 41, 42, 44, 34, 41, 28
C	39, 34, 39, 34, 46, 40, 37, 43, 35, 34, 34, 36, 25, 36, 50, 42, 42, 33, 40, 39, 24, 37, 31, 44, 45, 39, 26, 36, 27, 33, 40, 41, 44, 29, 45, 29
D	37, 33, 41, 39, 37, 36, 30, 33, 34, 34, 43, 39, 42, 39, 42, 33, 38, 38, 42, 30, 33, 41, 37, 39, 33, 41, 34, 34, 32, 37, 48, 40, 39, 42, 37, 30

1.4.2 STRIPCHARTS, STEM-AND-LEAF DISPLAYS, AND HISTOGRAMS

Before performing any calculations on the data we create a stripchart, a stem-and-leaf display, and a histogram of our data-set.

Stripchart it *the* most basic type of plot. It is obtained by plotting our points in order along a line. It is helpful to spot macroscopic errors in the data-set presented at a glance (wrong format of data, samples that are very different from the others, etc.), and also to have a visual clue of the spreading of the data in the exam, and consequently of the data location measurements presented later in this chapter (mean, variance, standard deviation, etc.).

The data in the exam does not show a *systematic* time bias, and so we can now continue the first visual inspection of the data by means of other tools.

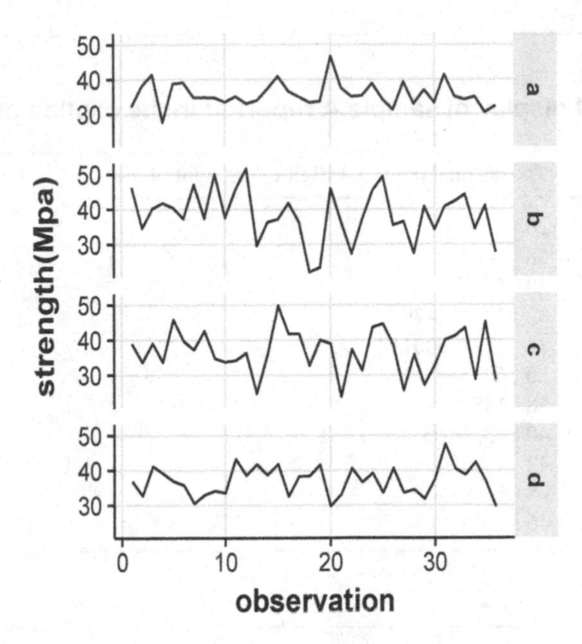

Figure 1.1 Stripchart of the tensile strengths (Mpa) for each series of experiments performed.

We draw then a *stem-and-leaf* display. It is a very simple plot to create that has the advantage to keep the original values of the measurements.

It is required to split each of them in a split *stem* and a *leaf*. For a set based on series A that contains the following values- 26.6, 30.5, 321.2, considering stem values of 26, 28, and 30, we have:

26.6 ⟶ 26 | 6
30.5 ⟶ 30 | 5
31.2 ⟶ 30 | 2

that can be written as:

26 | 6
28 |
30 | 52

Depending on the data, we can also select a different number of digits for the *stem* and for the *leaf*, i.e., for the first measurement of B, we have 22.06, 23.36, and 27.42, considering stem value of 140, written as 2|, we have:

2 | 23
2 | 778

Table 1.5
Stem-and-Leaf display of sample A reported in the caption of Table 1.4.

```
The decimal point is 1 digit to the right of the |

26 | 6
28 |
30 | 52
32 | 239034479
34 | 22778000112
36 | 6935
38 | 17803
40 | 834
42 |
44 |
46 | 7
```

Table 1.6
Stem-and-Leaf display of sample B reported in the caption of Table 1.4.

```
The decimal point is 1 digit(s) to the right of the |

2 | 23
2 | 778
3 | 044
3 | 55666677778
4 | 001112224
4 | 566679
5 | 02
```

Histograms are created in the following way. We calculate different *classes* with fixed *limits* dividing the span of the values measured by a fixed number of *intervals* n, and then we calculate the frequencies of the observation made according to each class. The number n is generally chosen between 5 and 20, depending on how many observations you measured.

Both stem-and-leaf and the histogram help us spot that the measurements performed by the different labs are *distributed* differently.

Labs A and D seem more *precise*, or in other words, the *distributions* of the value are smaller in comparison with labs B and C.

We can see that the *shapes* of the *distributions* of the observations are similar for labs A and D, and also for labs B and C.

Table 1.7
Stem-and-Leaf display of sample C reported in the caption of Table 1.4.

The decimal point is 1 digit(s) to the right of the —

```
2 | 4
2 | 56799
3 | 1334444
3 | 5666779999
4 | 000122344
4 | 556
5 | 0
```

Table 1.8
Stem-and-Leaf display of sample D reported in the caption for Table 1.4.

The decimal point is 1 digit(s) to the right of the —

```
28 | 57
30 | 58
32 | 5791556
34 | 157
36 | 69905
38 | 3457712
40 | 4662688
42 | 33
44 |
46 | 5
```

We can also spot that we have the maximum frequencies of observations at different values for labs A and D.

In other words, a series of measurements made under the same conditions (i.e., each series of data collected by each laboratory) represented graphically is known as a *frequency distribution.*

In order to better spot the trends in our data we report a histogram obtained with an higher number of observations, n=*360*.

For analytical data, the values are often distributed symmetrically about the mean value, the most common being the *normal error* or *Gaussian distribution curve*.

We will later see in the chapter how the normal error curve is the basis of a number of statistical tests that can be applied to assess effects of indeterminate errors, compare values, and describe the levels of confidence in the results.

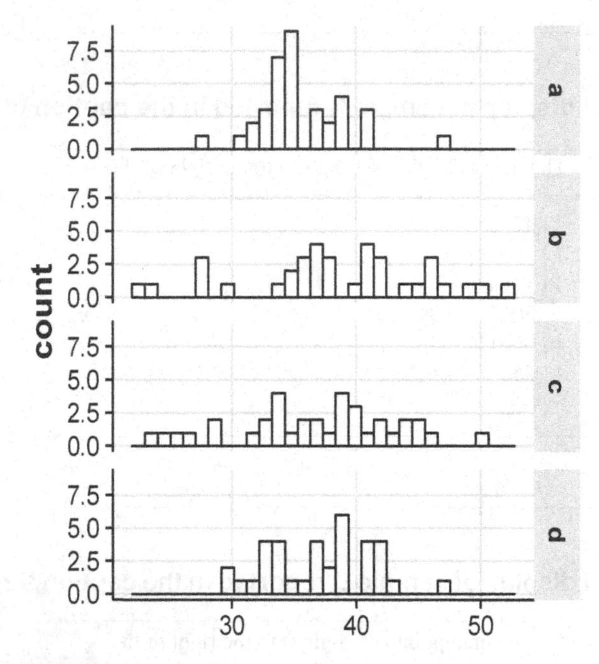

Figure 1.2 Histogram of the tensile strength (Mpa) for each series of experiments performed; n = 36.

We can compare the results obtained for our samples to a few common distributional shapes (see Figure 1.4).

Due to the sample size, the only observation that we can make is that we are not dealing with exponential or log normal distribution. It is also a safe assumption to exclude a *bimodal distribution* whole. We cannot rule out the other shapes of distributions presented.

It is often asked how many samples are necessary to obtain **reliable** conclusions. The question probably will have more meaning if we add to it which kind of test or calculation we would like to perform on our data.

It is possible indeed to calculate how many samples we will need to perform each calculation/test to obtain a result with a *desired confidence level.*

To calculate a mean, we only need two values, but the results obtained will probably not be very useful.

Also as we will see in Figure 1.4, each test will give different results depending on the number of observation performed. As a rule of thumb for the test introduced in this chapter, we should consider at least 10-12 measurements as a lower limit (13; 5).

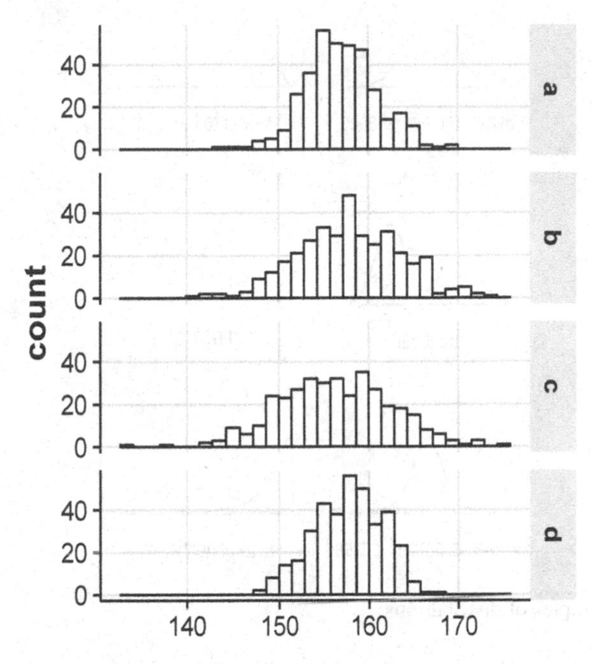

Figure 1.3 Histogram of the tensile strength (Mpa) for each series of experiments performed, with n = 360.

1.5 MODE, MEDIAN, MEAN, VARIANCE, AND STANDARD DEVIATION

We will now *quantify* the differences spotted among the different laboratories introducing numerical measurements calculated from our data, i.e., *descriptive statistics*.

In the following part of the chapter, we will introduce the most commonly used measure of center.

1.5.1 THE MEDIAN

Perhaps the simplest measure of the center of a data-set is the sample *median*. The *sample median* is the value that most nearly lies in the middle of the sample. It is the data value that splits the ordered data into two equal halves. To find the median, first arrange the observations in increasing order. In the array of ordered observations, the median is the middle value (if n is odd) or midway between the two middle values (if n is even). We denote the median of the sample by the symbol $\tilde{y}$ (read "y-tilde").

data set A sorted:

27.62, 30.48, 31.21, 32.17, 32.33, 32.91, 33.01, 33.26, 33.36, 33.44, 33.65, 33.89, 34.16, 34.23, 34.7, 34.74, 34.81, 34.96, 34.99, 35.04, 35.09, 35.1, 35.24, 36.56, 36.94, 37.26, 37.54, 38.11, 38.72, 38.84, 39.02, 39.3, 40.84, 41.34, 41.41, 46.66.

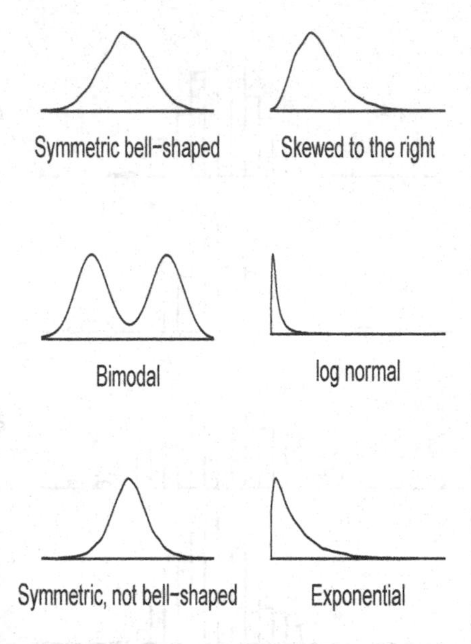

Figure 1.4 Examples of distributions.

Median calculated when the length of the data-set is an *odd* number, as in our case:

27.62, 30.48, 31.21, 32.17, 32.33, 32.91, 33.01, 33.26, 33.36, 33.44, 33.65, 33.89, 34.16, 34.23, 34.7, 34.74, 34.81, 34.96, **34.96**, 34.99, 35.04, 35.09, 35.1, 35.24, 36.56, 36.94, 37.26, 37.54, 38.11, 38.72, 38.84, 39.02, 39.3, 40.84, 41.34, 41.41, 46.66.

In order to show the reader how to calculated the median of a dataset when its length is an even number, we added an observation to the date set A and then reperformed the sorting

data set A sorted

22.06, 23.36, 27.31, 27.42, 27.7, 29.67, 34.16, 34.35, 34.61, 35.42, 36.16, 36.29, 36.33, 36.46, 36.97, 37.2, 37.36, **34.99**, **35.04**, 35.09, 35.1, 35.24, 35.64, 36.56, 36.94, 37.26, 37.54, 38.11, 38.72, 38.84, 39.02, 39.3, 40.84, 41.34, 41.41, 46.66

For all the labs, we have:

$$median\ series\ A = 32.20$$

$$median\ series\ B = 46.22$$

$$median\ series\ C = 38.93$$

$$median\ series\ D = 36.83$$

1.5.2 MODE

The mode can be defined as the most common value that occurs in a discrete data-set. If the set does not contain discrete values, we can group the data in classes in order to create classes and calculate it.

The mode for our labs

$$mode\ series\ A = 34.97$$

$$mode\ series\ B = 37.53$$

$$mode\ series\ C = 37.29$$

$$mode\ series\ D = 37.25$$

1.5.3 MEAN

Mean is calculated as sum of all measurements divided by the number of measurements. The *sample mean* is defined as:

$$\sum_{i=1}^{n} x_i/n \tag{1.1}$$

where the numerator refers to the sum of all the observations in the sample:

$$mean\ series\ A = \frac{x_1 + x_2 + x_3 + x_4 + \cdots}{n} =$$

$$\frac{32.17 + 38.11 + 41.34 + 27.62 + \cdots}{36} = \frac{1282.91}{36} = 35.64$$

and for the other laboratories we will have:

$$mean\ series\ B = \frac{46.23 + 34.61 + 40.18 + 41.8 + \cdots}{36} = \frac{1382.59}{36} = 38.41$$

$$mean\ series\ C = \frac{38.94 + 33.61 + 39.07 + 33.61 + \cdots}{36} = \frac{1322.99}{36} = 36.75$$

$$mean\ series\ D = \frac{36.87 + 32.75 + 41.18 + 39.08 + \cdots}{36} = \frac{1334.53}{36} = 37.07$$

The information gathered till now can give a hint that the center of the *distributions* of the measure are different and also their span, but we do not have enough elements for a sound proof of it.

Resistance when a descriptive statistics is relatively unaffected by the changes of a small portion of the data, even if the changes are large, is called *resistant*.

We have considered the shapes and centers of distributions, but a good description of a distribution should also characterize how spread out the distribution is the range, the standard deviation, and the coefficient of variation (16; 15).

1.5.4 A VISUAL COMPARISON OF MEAN, MEDIAN, AND MODE

We introduced the concept of distribution before in the text. Mean, median, and mode can give hints about our data distribution. The following figure will demonstrate why it is useful to calculate mean, median, and mode.

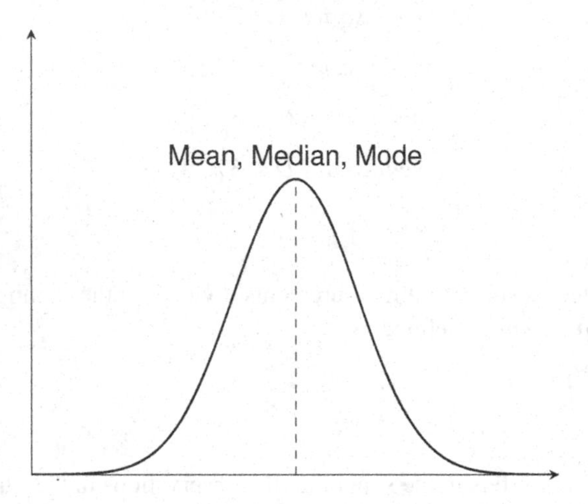

Figure 1.5 Mean, median, and mode in a symmetric distribution.

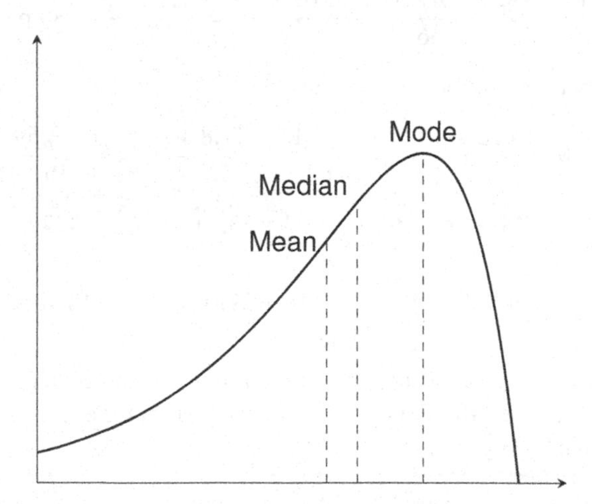

Figure 1.6 Mean, median, and mode in a skewed distribution.

As you can see, when we are working with a *normal distribution*, all the three measurements of location have the same value, while when our data follows a *skewed* distribution, this is not true.

1.5.5 THE RANGE

The sample range is the difference between the largest and smallest observations in a sample.

$$range(x) = max(x) - min(x) \tag{1.2}$$

Table 1.9
Ranges for laboratories A, B, C, and D.

	A	B	C	D
min(x)	27.617	22.061	23.816	29.545
max(x)	46.663	51.720	49.869	47.531

1.5.6 QUARTILE AND INTERQUARTILE RANGE

When a set of n measurements of the variable x has been arranged in order of magnitude, the pth percentile is the value of x greater than p of the measurements

The 25th and 75th percentile are called the lower and upper quartiles, and the 50th percentile is the median of our data-set. The interquartile range (IQR) for a set of measurements is the difference between the upper and lower quartile.

In order to summarize our data, we can report our five-number summary that consists of the minimum, lower quartile, the median, the upper quartile, and finally the maximum.

Rank position of quartiles are:

First quartile (.25)(n+1),

Second quartile (.50)(n+1),

Third quartile (.70)(n+1).

Table 1.10
Quartiles for laboratories A, B, C, and D.

	qA	qB	qC	qD
0%	27.617	22.061	23.816	29.545
25%	33.419	35.216	33.393	33.512
50%	34.975	37.528	37.291	37.253
75%	37.680	42.724	41.309	40.446
100%	46.663	51.720	49.869	47.531

1.5.7 VARIANCE

The *variance of a population*, σ^2, of N measurement is the average of the squares of the deviation of the measurements (X) from their mean μ:

$$\sigma^2 = \frac{\sum_{i=1}^{N} (X-\mu)^2}{N} \tag{1.3}$$

Analogously, the *variance of a sample, s^2*, of n measurement is the average of the squares of the deviation of the measurements from their mean.

$$s^2 = \frac{\sum_{i=1}^{n} (X-\overline{X})^2}{n-1} \tag{1.4}$$

$$s^2 \; series\, A = \frac{(x_1-\overline{x})^2+(x_2-\overline{x})^2+(x_3-\overline{x})^2+(x_4-\overline{x})^2+\cdots}{36-1}$$

$$s^2 \; series\, A = \frac{(32.17-35.64)^2+(38.11-35.64)^2+(41.34-35.64)^2+\cdots}{36-1} = 12.94$$

$$s^2 \; series\, B = \frac{(46.23-38.41)^2+(34.61-38.41)^2+(40.18-38.41)^2+\cdots}{36-1} = 52.46$$

$$s^2 \; series\, C = \frac{(38.94-36.75)^2+(33.61-36.75)^2+(39.07-36.75)^2+\cdots}{36-1} = 40.02$$

$$s^2 \; series\, D = \frac{(36.87-37.07)^2+(32.75-37.07)^2+(41.18-37.07)^2+\cdots}{36-1} = 17.78$$

Table 1.11
Variances for laboratories A, B, C, and D for all calculated digits.

A	B	C	D
12.94	52.46	40.02	17.78

Degree of freedom: It is not easy to define the degrees of freedom, since they have a different meaning depending on the subject you are taking in the exam. As a chemist, I had at first gotten familiar with degrees of freedom while I was studying physical chemistry, and so their definition (F) involved the parameters related to the number of phases (P) and the number of components C in a heterogeneous system existing in equilibrium, or more in general $F = C+2-P$. In statistics, tests generally are defined as the number of values involved in the final calculation of statistics that are free to vary. As a practical example, if you have 3 measurements, since you can calculate their mean, you have 2 degrees of freedom.

1.5.8 THE STANDARD DEVIATION

The standard deviation is the classical and most widely used measure of dispersion. Recall that a deviation is the difference between an observation and the sample mean:

$$s = \sqrt{\frac{\sum_{i=1}^{n}(x_i - \overline{x})^2}{N-1}} \tag{1.5}$$

where $\sum_{i=1}^{n}(x_i - \overline{x})^2$ represents the sum of the squared deviations.

Analogously to what we reported for the variance, the calculations performed on the data-set are the following:

$$s\ series\ A = \sqrt{\frac{(x_1 - \overline{x})^2 + (x_2 - \overline{x})^2 + (x_3 - \overline{x})^2 + (x_4 - \overline{x})^2 + \cdots}{N-1}}$$

$$s\ series\ A = \sqrt{\frac{(32.17 - 35.64)^2 + (38.11 - 35.64)^2 + (38.11 - 35.64)^2 + \cdots}{35}}$$

Table 1.12
Standard deviations for laboratories A, B, C, and D.

A	B	C	D
3.598	7.243	6.326	4.217

Standard deviation and variance for Labs A, B, C, and D confirm that the span of the measurement is different. We still need a tool for quantifying this difference.

1.5.9 DISTRIBUTIONS

We've seen that when we plot our data we can have information about their *frequency distribution.*

After introducing the mean, median, and standard deviation, we are now able to write distributions as their functions. This has a very useful practical aspect from our point of view. With just a few parameters, we can spot if our data is normally distributed or not, and use the more appropriate tools while dealing with them.

When data is *normally distributed*, the more measures we take of our population, the more the distribution probability density will resemble a bell-shaped curve. The normal distribution can be expressed in the following way:

$$f(x;\mu,\sigma^2) = \frac{1}{\sigma\sqrt{2\pi}} exp\left[-\frac{(x-\mu)^2}{2\sigma^2}\right] \tag{1.6}$$

where x, is the measurement performed on a sample, μ is the mean, and σ is the standard deviation. For the *standard normal distribution*, we have:

$$f(z) = \frac{1}{\sqrt{2\pi}} exp^{-z^2/2} \tag{1.7}$$

Why focusing on the standard normal distribution?

The standard normal distribution has several properties that renders it unique.

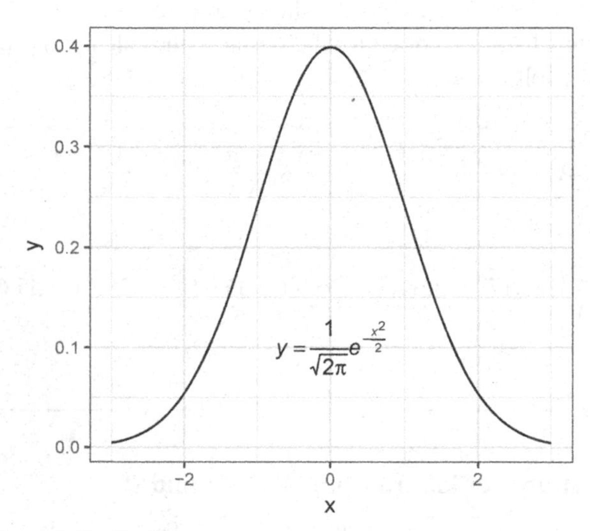

Figure 1.7 Normal distribution.

One of the most important aspect of the normal distribution is due to the central limit theorem (*CLT*) that states that the distribution of means, regardless of the shape of their parent distribution approach a normal distribution as the sample size increases. All the distributions related to it, such as t, F, and χ^2 share these properties and consequently can be used for testing means and function of means. A few suggestions about the use of the different distributions- for *discrete values*

- *Binomial* distribution is convenient to use only for small samples
- *Poisson* distribution is useful for studying an event with a small probability and many possibilities

and for *continuous values*:

- *z standard normal distribution*, with 50% of the observation on either side of the midpoint. The mean of the standard normal observation set to 0 and standard deviation to 1. The 95% of the area under the curve is located between -1.96 and +1.96 standard deviations
- *t-distributions* (note the plural since there is a family of them) similar to normal distribution. The most commonly used statistical distributions. The

number of observations and so the degrees of freedom determine which of the t-distribution to use
- χ^2 distributions are skewed, with value starting at 0
- F distributions also skewed, with values starting at 0

All these distributions are derived from the *normal* distribution, and in order to make use of their properties and perform any kind of test on our data-set, we should check them prior for *normality*.

1.6 Z-SCORE

Sometimes, we need to know the position of one observation relative to others in a set of data. Mean and standard deviation can be used to calculate z-score, which measures the relative standing of a measurement in a data-set, and it is defined as:

$$z_i = \frac{(x_i - \overline{x})^2}{s} \tag{1.8}$$

i.e., z. score measures the distance between an observation and the mean, measured in units of sample standard deviation. In Table 1.13 we report the summary of the data-set for case study 1.1:

1.6.1 BOX AND WHISKERS PLOT

After introducing several descriptive statistics, we will show how we can include them in a visual representation of the data. A box plot can be defined as the simplest graph for quantitative data analysis where you can plot the measurements as a point in an horizontal axis. It works well with a small data-set.

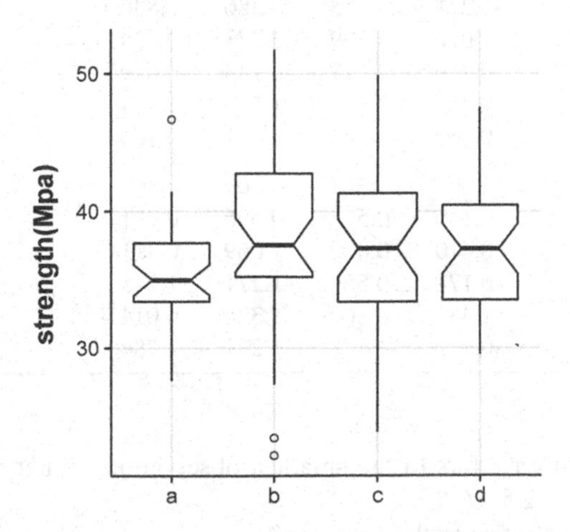

Figure 1.8 Box plot tensile strength (Mpa) for each series of experiments performed.

Table 1.13
Z Score for Labs A, B, C, and D.

A	B	C	D
-0.963	1.080	0.346	-0.047
0.688	-0.524	-0.496	-1.025
1.585	0.245	0.366	0.975
-2.229	0.468	-0.497	0.477
0.856	0.279	1.436	-0.043
0.942	-0.144	0.471	-0.316
-0.260	1.196	0.074	-1.561
-0.229	-0.134	0.933	-0.981
-0.249	1.616	-0.314	-0.697
-0.611	-0.108	-0.478	-0.843
-0.152	1.025	-0.407	1.478
-0.731	1.838	-0.067	0.348
-0.484	-1.206	-1.902	1.111
0.451	-0.287	-0.151	0.384
1.446	-0.166	2.074	1.129
0.256	0.471	0.804	-1.072
-0.189	-0.291	0.796	0.285
-0.634	-2.257	-0.635	0.316
-0.552	-2.078	0.518	1.075
3.065	1.038	0.356	-1.747
0.528	-0.310	-2.044	-0.945
-0.167	-1.532	0.098	0.830
-0.111	-0.198	-0.861	-0.121
0.890	0.950	1.089	0.499
-0.392	1.514	1.245	-0.847
-1.231	-0.413	0.286	0.830
1.018	-0.269	-1.773	-0.829
-0.759	-1.517	-0.143	-0.619
0.362	0.325	-1.530	-1.254
-0.662	-0.586	-0.632	0.100
1.605	0.321	0.509	2.481
-0.150	0.528	0.696	0.791
-0.410	0.802	1.069	0.391
-0.178	-0.560	-1.271	1.245
-1.432	0.362	1.326	-0.014
-0.919	-1.478	-1.291	-1.784

- The lower whisker refers to the smallest observation greater than or equal to $lowerhinge - 1.5 \cdot IQR$
- The lower hinge refers to the 1st quartile
- The lower edge of notch to the median - $1.58 \cdot IQR/sqrt(n)$

- The middle represents the median
- The upper edge of notch to the median + $1.58 \cdot IQR/sqrt(n)$,
- The upper hinge to the 3rd quartile
- The upper whisker represents the largest observation less than or equal to the upper hinge + $1.5 \cdot IQR$
- The asterisks represent observations considered as outliers (outside the value of the hinges $\pm 1.5 \cdot IQR$)

In this way, we can summarize and highlight better the results gathered till now for the four labs. Labs A and D are the most precise, since the height of the box and whiskers is shorter in comparison with the others. B and C are less precise (longer box and whiskers). D, B, and C are the most accurate, since the mean of the observation gathered is near the "true" value of 38. We can also spot the presence of one candidate outlier for set A, and two candidate outliers for set B.

1.7 ERROR PROPAGATION AND UNCERTAINTY

The definition of standard deviation and the concept of distributions are necessary for introducing the basic concepts of *error propagation.* As reported by (10), the procedures for combining *random* and *systematic* errors previously defined are different. Starting with *random errors*, when a value of y, is calculated from a linear combination of measured quantities a, b, c, by:

$$y = k + k_a a + k_b b + k_c c + \cdots \quad (1.9)$$

where k, k_a, k_b, K_c, etc., are constants and for the standard deviations we have that:

$$\sigma_y = \sqrt{(k_a \sigma_a)^2 + (k_b \sigma_b)^2 + (k_c \sigma_c)^2 \cdots} \quad (1.10)$$

If we calculate y from a multiplicative expression as:

$$y = kab/cd \quad (1.11)$$

where a, b, c are **independent**, the relationship between the squares of the *relative* standard deviations is:

$$\frac{\sigma_y}{y} = \sqrt{\left(\frac{\sigma_a}{a}\right)^2 + \left(\frac{\sigma_b}{b}\right)^2 + \left(\frac{\sigma_c}{c}\right)^2 + \cdots} \quad (1.12)$$

When the quantities are **not independent** (i.e., a quantity raised to power $y = b^n$) the standard deviations of y and b are related by:

$$\left|\frac{\sigma_y}{y}\right| = \left|\frac{n\sigma_b}{b}\right| \quad (1.13)$$

In general for $x, y = f(x)$ the standard relations are related by:

$$\sigma_y = \left|\sigma_x \frac{dy}{dx}\right| \quad (1.14)$$

The rules for calculating *systematic errors* are the following:
Linear combinations:

$$\Delta y = k + k_a \Delta a + k_b \Delta b + k_c \delta c + \cdots \tag{1.15}$$

where $\Delta a, \Delta b, \Delta K_c$ are the systematic errors.
Multiplicative expression:

$$\left(\frac{\Delta y}{y}\right) = \left(\frac{\Delta a}{a}\right) + \left(\frac{\Delta b}{b}\right) + \left(\frac{\Delta c}{c}\right) + \cdots \tag{1.16}$$

quantity raised to some power:

$$\frac{\Delta_y}{y} = \frac{n\Delta_b}{b} \tag{1.17}$$

Both random and systematic errors contribute to an estimated combined error referred to as *uncertainty*. The Guide to the Expression of Uncertainty in Measurement (GUM) defines *uncertainty* as "a parameter, associated with the result of a measurement, that characterizes the dispersion of the values that could reasonably be attributed to the measurand" (4).

We can further investigate the concept of uncertainty and define also:

- *Standard uncertainty* (u) that refers to the concept of standard deviation
- (U) a *expanded uncertainty* obtained by $U = uxk$, where K is a coverage factor with a default value of 2 that corresponds to a 95% interval

Both *bottom up* and *top-down* approach can be employed to estimate uncertainty. *Bottom up* approach tries to *identify* and *separate* each stage of the analysis and determine the associated systematic and random error, and then combines then according to the rules we just presented.

In order then to combine together random and systematic errors, the latter ones are considered coming from a uniform distribution. Their contribution is obtained by dividing the error by $\sqrt{3}$ (rectangular distribution) or $\sqrt{6}$ (triangular distribution). These errors are referred to as Type B error (while random errors previously presented are referred to as Type A errors).

Again with this approach we assume that the sources of error are all **independent** from each other.

Among the contribution to uncertainty it is useful to define also a *combined standard uncertainty* defined as:

> the result of a measurement when that result is obtained from the values of a number of other quantities, equal to the positive square root of equal to the positive square root of a sum of terms, the terms being the variances or covariances of these other quantities weighted according to how the measurement result varies with changes in these quantities.

Provided a set of quantities we can then obtain the uncertainty of a derived measurement as *propagation of uncertainty*. The two main methods as reported in (18) are the Taylor series method (TSM) and the Monte Carlo Method. TSM is based on a Taylor expansion of the mathematical expression that produces the output variables, while Monte Carlo method deals with generalized input distributions and propagates the uncertainty via Monte Carlo simulation.

When we are not able to split every error contribution, and when it is possible to repeat the experiment in other laboratories, it is possible to recur to a *top down* approach. It is based on using results from *proficiency testing schemes (PT)*, where aliquots of homogeneous materials are circulated to a number of laboratories, analyzed with a scheduled frequency, and the results obtained are reported to a central laboratory. Due to its peculiarity this kind of workflow can be applied only in selected fields.

These principles are studied to reproduce as much as possible the conditions of everyday laboratory practice, even if in some cases, it may be inadequate.

For further details, readers can refer to (4).

1.8 NORMALITY TESTS

Statistical tests that include t-test and analysis of variance assume that our data follows a *normal* or *Gaussian distribution*, o before applying them to our data and avoid misinterpreting the results, we need to check this assumption. We already visually checked our data by means of histograms, but we can also employ *Q-Q plot* or *significance test*

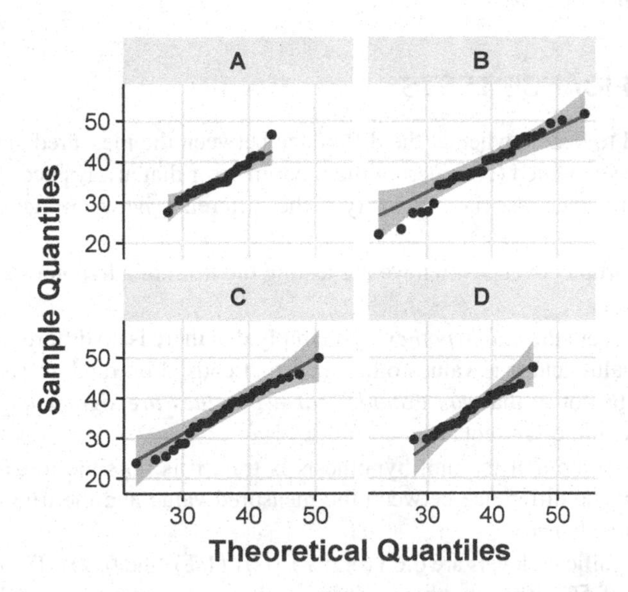

Figure 1.9 Q-Q plot of the tensile strength (Mpa) for each series of experiments performed.

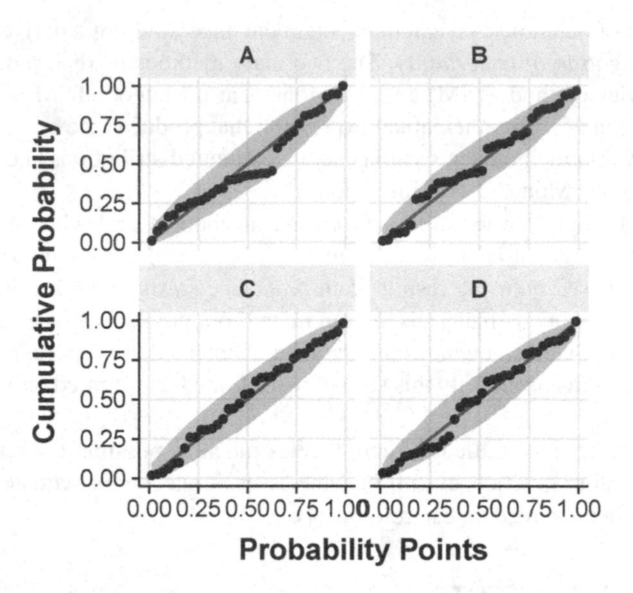

Figure 1.10 P-P plot of the tensile strength (Mpa) for each series of experiments performed.

Plot 1.10 highlights even better in comparison with the previous box and whiskers plot the shorter span of Labs A and D, and also the reduced confidence interval (gray areas near the points). P-P plot shows that all distributions can be considered *normal* with a confidence of 95%.

1.9 SIGNIFICANCE TESTS

When we need to decide whether the difference between the measured and standard amount can be accounted (considering the random error that always occurs), we can employ a significance test (i.e., to verify if the difference between them is significant).

A very important concept when we are testing the truth of a hypothesis is the *null hypothesis (H0)*.

When you accept the *null hypothesis* you imply that there is no difference between the observed value and true values other than that caused by random variation. So it's important to notice that *you exclude that two values are significantly different* that doesn't mean they are equal.

When we assume that the null hypothesis is true, it is possible to calculate the probability that the difference between the measured value and the true values is a result of random chance.

Common significant levels are 0.05 (or 5%), 0.01 (1%), and 0.001. If we calculate a significance of 5%, it means that we have a chance to reject the null hypothesis when it is true. Analogously for 0.01 and 0.001, it means that we reject the null hypothesis when we have the chance of 1 in 100 (and 1 in 1000) to be correct.

When we need to establish whether *one experimental value* is significantly greater than another, or the other way around, we recur to a *one-tailed* test, while when we need to establish whether there is a significant difference between the *two values* being compared, we make use of the *two-tailed* test.

1.9.1 OUTLIERS

We will also briefly introduce the concept of an outlier. It's a common task to inspect a set of replicate measurements, and find a value which appears to be outside the range expected, even considering indeterminate random errors alone. These values are termed *suspect values* or *outliers*. Sometimes, the recourse to statistical tools is not necessary, since the cause of anomaly can be trivially discovered. However, if this is not the case, we have different tools we can make use of.

1.9.2 Q-TEST

You can apply it after checking that your data is normally distributed. Also knows as Dixon's Q-test, it is based on the calculation of a ratio Q_{exptl} defined as the absolute difference between a suspect value and its closest value divided by the spread of all the values in the set:

$$Q_{exptl} = |suspectvalue - nearestvalue| / (largestvalue - smallestvalue)$$

The value obtained is then compared with tabulated values (Q_{tab}) at selected levels of probability for a set of n values.

If Q_{exptl} is less than Q_{tab} then the null hypothesis (no significant difference) is accepted, and the value is retained for the whole data processing, otherwise the value is rejected and it should not be used in the calculations. We will use the Q-test on a smaller data-set based on lab A measurements, since it should not be used for samples > 30.

Table 1.14
Results for Dixon test performed on the data from Lab A.

Statistic	P-value	Method	Alternative
0.367	0.116	Dixon test for outliers	highest value 46.66 is an outlier

1.9.3 COCHRAN TEST

As we've seen for Dixon's test, Cochran test can be applied after checking that data is normally distributed. It is used to test for outlying or inlying variance. It is used to test if the largest variance in several *groups* of data is outlying and we should reject this group.

Table 1.15
Results for Dixon test performed on the data from the lab, including a mislabeled sample in the data-set. Note the difference in the p-value in comparison with the previous case.

Statistic	P-value	Method	Alternative
0.454	0.021	Dixon test for outliers	highest value 49.098 is an outlier

Table 1.16
Results for Cochran for data-set gathered from Labs A, B, C, and D.

P-value	Method	Alternative
$2 \cdot 10^{-16}$	Cochran test for outlying variance	Group C has outlying variance

1.10 T-TEST

It is used to compare the experimental means of two sets of data or to compare the experimental mean of one set of data with a known "true" value. Depending on its application we need to apply the following equations:

$$t = \frac{(\overline{x}_A - \overline{x}_B)}{S_{pooled}} \times (\frac{NM}{N+M})^{1/2} \tag{1.18}$$

where s_{pooled} is the pooled estimated standard deviation for sets A and B, and N and M are the numbers of values in sets A and B, respectively. s_{pooled} is calculated as:

$$s_{pooled} = \sqrt{\frac{(N-1)s_A^2 + (M-1)s_B^2}{(N+M-2)}} \tag{1.19}$$

In some circumstances this equation may not be appropriate (i.e. when we have only one specimen for each set or when we previously know that the specimen are not characterized in the same experimental condition) then we need to relay on a paired t-test where t is defined as:) that we need to reply on a *paired t-test* where t is defined by:

$$t = \frac{(\overline{x}_d)}{S_d} \times N^{1/2} \tag{1.20}$$

where $\overline{x_D}$ is the mean difference between paired values and s_D is the estimated standard deviation of the differences. When we compare one experimental mean with a known value μ we use:

$$t = \frac{(\overline{x} - \mu)}{s} \times N^{1/2} \tag{1.21}$$

Using the appropriate formula, t_{expl} is calculated. If its value is less than t_{tan} we accept the null hypothesis, i.e., there is no evidence of a bias and no significant difference between the experimental mean and a known value, otherwise the null hypothesis is rejected. Two tailed t-tests are generally preferred, but both one-tailed and two-tailed versions of the t-test are available.

We report how to calculate t-tests for the pair A and B, and then the summary for all the pairs in the data-set analyzed is:

$$s_{pooled} = \sqrt{\frac{(36-1)\times 3.6^2 + (36-1)\times 7.24^2}{(36+36-2)}} =$$

$$\sqrt{\frac{(35)\times 12.94 + (35)\times 52.46}{(70)}} = \sqrt{\frac{(452.97)+(1836.15)}{(70)}} =$$

$$\sqrt{\frac{(2289.12)}{(70)}} = \sqrt{(32.7)} = 5.72$$

and with a t_{exptl} value calculated as:

$$t_{exptl} = \frac{(35.64)-(38.41)(5.72)}{\times}(\frac{1296}{36+36})^{1/2}$$

$$t_{exptl} = \frac{(-2.77)(5.72)}{\times}(\frac{1296}{72})^{1/2}$$

$$t_{exptl} = (-0.48)\times(4.24) = -2.05$$

Table 1.17
T-test for pairs A-B, A-C, A-D, B-C, and B-D.

	Estimate	Statistic	P-value	Conf.low	Conf.high
AB	**-2.769**	**-2.054**	**0.045**	**-5.475**	**-0.063**
A-C	-1.113	-0.918	0.363	-3.544	1.317
A-D	-1.434	-1.552	0.125	-3.277	0.409
B-C	1.656	1.033	0.305	-1.542	4.853
B-D	1.335	0.956	0.343	-1.463	4.133
C-D	-0.321	-0.253	0.801	-2.854	2.213

For our data-set, we notice that the only p-value lower than 0.1 is ascribed by the AB pair, while for the other we have higher values, implicating that we cannot find a statistically significant difference for the series in the exam.

After introducing the concept of *normal distribution*, t, t_{exptl}, and standard deviation, we can calculate if we would like to estimate the mean of a sample with an error estimation less than a δ with a probability of 95%. If the standard deviation is known, we can write that:

$$\delta = \frac{\sigma}{\sqrt{N}} z_{1-0.025}$$

$$N \geq \left(\frac{1.96}{\delta}\right)^2 \sigma^2$$

$$N = (z_{1-\alpha/2} + z_{1-\beta})^2 \left(\frac{\sigma}{\delta}\right)^2 \rightarrow two-sided\ test$$

$$N = (z_{1-\alpha} + z_{1-\beta})^2 \left(\frac{\sigma}{\delta}\right)^2 \rightarrow one-sided\ test$$

If we do know the standard deviation, we must compute the sample size with the population standard deviation using the following equations:

$$N = (t_{1-\alpha/2} + t_{1-\beta})^2 \left(\frac{s}{\delta}\right)^2 \rightarrow two-sided\ test$$

$$N = (t_{1-\alpha} + t_{1-\beta})^2 \left(\frac{s}{\delta}\right)^2 \rightarrow one-sided\ test$$

Since t critical values of the t distribution depend on the known number of measurements (and degrees of freedom) we are trying to estimate, we need an initial guess and then to iterate the results. Tabulated values already exist, showing minimum sample sizes for a two-sided test. For a value of $\delta = 1\sigma$, $\alpha = 0.05$ and $\beta = 0.10$, the minimum value is **N = 11**.

1.11 F-TEST

It is defined as the ratio of the population variances σ_1^2/σ_2^2, or the samples variances ${s_1}^2/s_2^2$, of two data-sets, where the larger variance is always placed in the numerator, so that $F \geq 1$.

If the *null hypothesis* is true, the variances are equal and the value of F will be close or equal to 1. An experimental value of F_{exptl} is calculated and compared with tabulated values, determined for defined probability level (usually $90-95\%$) and for a number of degrees of freedom ($N-1$) for each set of data.

If F_{exptl} is less than F_{tab}, then we accept the null hypothesis that there is no significant difference between the two variances and consequently between the precision of the data-set, otherwise the null hypothesis is rejected.

In this case it can be useful to perform the test as a *one-tailed F-TEST* in order to show whether the precision of one set of data is significantly better than the other, or to perform the test as a *two-tailed F-Test* to spot if the precision of the two data-sets are significantly different.

F-Test is generally used to compare the precision of two data-sets that are performed by two analysts in the same laboratory, two different methods of analysis, or by two different laboratories.

We report the calculations for the pairs in the selected data-set:

$$F_i = \frac{s_1^2}{s_2^2}$$

$$F_{AC} = \frac{s_a^2}{s_b^2} = \frac{12.942}{52.461} = 0.247$$

$$F_{AC} = \frac{s_a^2}{s_c^2} = \frac{12.942}{40.024} = 0.323$$

$$F_{AD} = \frac{s_a^2}{s_d^2} = \frac{12.942}{17.783} = 0.728$$

$$F_{BC} = \frac{s_b^2}{s_c^2} = \frac{52.461}{40.024} = 1.311$$

$$F_{BD} = \frac{s_b^2}{s_d^2} = \frac{52.461}{17.783} = 2.95$$

Table 1.18
F-test for pairs A-B, A-C, A-D, B-C, B-D, and C-D.

	Estimate	Statistic	P-value	Conf.low	Conf.high
A-B	0.247	0.247	0.000	0.126	0.484
A-C	0.323	0.323	0.001	0.165	0.634
A-D	0.728	0.728	0.352	0.371	1.427
B-C	1.311	1.311	0.427	0.668	2.571
B-D	2.950	2.950	0.002	1.504	5.785
A-C	2.251	2.251	0.019	1.148	4.414

We have a value greater than the significant level of 0.05 only for pairs A-D and B-C, which means that there is no significant difference among these pairs.

1.12 ONE-WAY ANALYSIS OF VARIANCE ANOVA

One-way analysis of variance (one-way ANOVA) is an extension of the *t-test* for *independent* samples where we need to compare more than two groups. The *null hypothesis* will be that the means of the different groups are the same, while the *alternative hypothesis* is that at least one sample mean is not equal to the others.

When *p*-value is less than a predefined value, we assess that there is no difference among the groups in the model summary. If the *p*-value is higher, we can apply a $Tukeypairwise - comparison$ to evaluate the p between specific pairs.

Table 1.19
Summarized T-test results for data-set presented in Table 1.3.

Term	Df	Sumsq	Meansq	Statistic	P-value
variable	3	140.310	46.770	1.518	0.212
Residuals	140	4312.359	30.803	NA	NA

Table 1.20
Summary of ANOVA performed on tensile strength for the data set in exam reported in 1.3.

Term	Comparison	Estimate	Conf.low	Conf.high	Adj.p-value
variable	B-A	2.769	-0.632	6.170	0.153
variable	C-A	1.113	-2.288	4.515	0.830
variable	D-A	1.434	-1.967	4.835	0.692
variable	C-A	-1.656	-5.057	1.746	0.586
variable	D-B	-1.335	-4.737	2.066	0.738
variable	D-C	0.321	-3.081	3.722	0.995

1.13 TWO-WAY ANALYSIS OF VARIANCE ANOVA

Two-way Analysis of variance is used when we need to assess the effects of two grouping variables on a response variable. In this case, the test *hypothesis* are:

- NO difference in the means of first factor
- NO difference in the means of second factor
- NO interactions between factors

The *alternative hypothesis* for the first two cases is that the means are not equal, while the alternative hypothesis for the third case is that there is no interaction between the first and second factor. In order to show an example, we will rely on a data-set based on data presented in Table 1.2, reported in Table 1.21.

In this case we have applied the two-way ANOVA to two samples from the data-set reported in Table 1.21. The data gathered refers to the hardness of the two samples of different concentrations of GO and two samples of different concentrations of FGO. As we can read in the results table, the *p-value* for the kind of *additive is not significant* (0.076), while the *concentration is significant* (2e-16), so indeed a very low value.

1.13.1 TWO WAY ANOVA WITH INTERACTION

Testing a model where we considered interactions between the factors, we saw that it was statistically significant. Does this result make sense from a practical point of view?

Table 1.21
Data-set based on (3).

GO c=0.25	GO c=0.5	FGO c=0.25	FGO c=0.5
147.586	174.669	157.511	176.126
150.555	173.467	149.622	175.571
152.169	172.266	148.453	174.992
145.309	172.488	149.494	177.905
150.858	182.248	143.855	178.013
151.012	175.402	146.162	172.776
148.851	173.528	141.100	177.672
148.907	173.678	145.295	175.564
148.871	176.379	150.529	175.940
148.220	172.919	149.671	179.351
149.046	170.655	159.247	182.252
148.003	176.724	146.657	186.591
148.447	171.929	147.723	176.907
150.129	174.955	150.597	186.424
151.919	172.192	147.028	175.369

Table 1.22
ANOVA for data-set reported in Table 1.21.

Term	Df	Sumsq	Meansq	Statistic	P-value
label.additive	1	43.372	43.372	3.256	0.076
label.conc	1	10992.517	10992.517	825.343	2e-16
Residuals	57	759.167	13.319	NA	NA

Table 1.23
T-test for two independent variables.

Lhs	Rhs	Estimate	Std.error	Statistic	P-value
0.5-0.25	0	27.071	0.942	28.729	2e-16

We can interpret that with a concentration of additive of 0.5 of FGO and GO, we can spot a difference between the samples that is not present with a 0.25 conc. of additive.

This can be a hint of a different mechanism in the effect of the additive, and would be interesting to investigate further by gathering data with different techniques, and eventually to also measure other properties not limited to hardness.

In order to also have a visual comparison of the effects of FGO and GO, we report a residual plot in Figure 1.12, and a Q-Q plot in Figure 1.11 for the data-set in the exam.

Both of them are also useful to spot the presence of outliers and to check the normality assumption on our data.

Residual plots highlight measurements 41, 31, and 57 as suspicious values. Q-Q plot shows that there are a few samples not normally distributed. How can we check *numerically*?

A check on the normality assumption can be performed employing the *Shapiro-Wilk* test, and in Table 1.26 we reported the results for the residuals. As it can be seen, they are normally distributed (the values are higher than the threshold for $p>0.01$) for each class, and we can confirm the anova performed.

Table 1.24
Summary of Two way ANOVA considering interaction.

Term	Df	Sumsq	Meansq	Statistic	P-value
label.a	1	43.372	43.372	3.525	0.066
label.conc	1	10992.517	10992.517	893.454	0.000
label.a:label.conc	1	70.177	70.177	5.704	0.020
Residuals	56	688.990	12.303	NA	NA

Table 1.25
Comparison between group for two way ANOVA interaction.

Term	Comparison	Estimate	Conf.low	Conf.high	Adj.p-value
label.a	GO-FGO	-1.700	-3.515	0.114	0.066
label.conc	0.5-0.25	27.071	25.257	28.885	0.000
label.a:label.conc	GO:0.25-FGO:0.25	0.463	-2.929	3.854	0.984
label.a:label.conc	FGO:0.5-FGO:0.25	29.234	25.842	32.625	0.000
label.a:label.conc	GO:0.5-FGO:0.25	25.370	21.979	28.762	0.000
label.a:label.conc	FGO:0.5-GO:0.25	28.771	25.380	32.163	0.000
label.a:label.conc	GO:0.5-GO:0.25	24.908	21.517	28.299	0.000
label.a:label.conc	GO:0.5-FGO:0.5	-3.863	-7.255	-0.472	0.020

Table 1.26
Shapiro-Wilk normality test.

Set	Statistic	P-value	Method
GO 0.25	0.95713	0.6226	Shapiro-Wilk normality test
GO 0.50	0.85731	0.02207	Shapiro-Wilk normality test
FGO 0.25C	0.91614	0.1681	Shapiro-Wilk normality test
FGO 0.50C	0.84407	0.0143	Shapiro-Wilk normality test

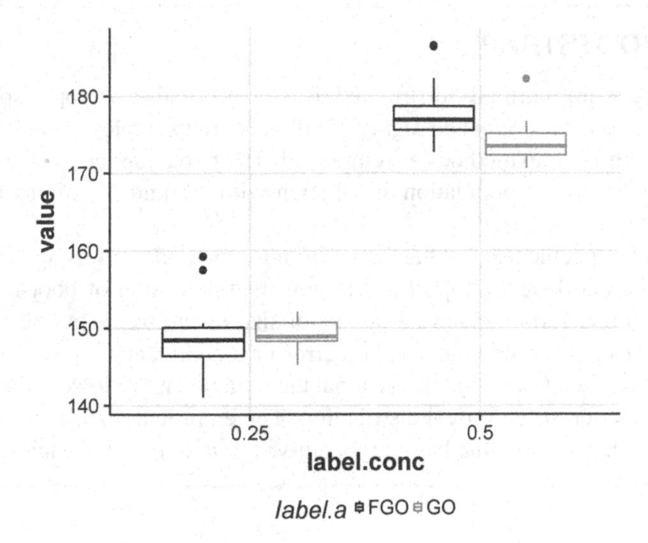

Figure 1.11 Two way analysis of variance.

1.14 TYPE I, II, AND III ERRORS

When the data-set is unbalanced (we do not have the same number of observations for each group), there are different methods for calculating the sum of squares of the residuals. SAS software called them Type I, II, and III.

- Type I is also called "sequential" sum of squares. Considering two effectors A and B, it tests the main effector A, followed by the main effect of B after the main effect of A, and finally the interaction of A and B after the main effect.
- Type II assumes no significant interactions, and is equivalent to compute type I analysis in reverse order of the factor.
- Type III tests the presence of a main effect after the other main effect and interaction.

1.15 BOOTSTRAP

If our data is not normally distributed, it is also possible to apply specific tests. *Wilcoxon Rank Sum* and *Signed Rank tests* are examples, and also randomization/bootstrap methods are employed. The bootstrap idea is simply to replace the unknown population distribution with the known empirical distribution.

We return to specific text for the theory of these methods and its application as (2; 7; 1), and here we report just an informal definition of bootstrap. In its most general form, we have a sample of size n, and we want to estimate a parameter or determine the standard error or a confidence interval for the parameter or even test a hypothesis about the parameter. We look at the sample and consider the empirical distribution. The empirical distribution is the probability distribution that has a probability, $1 = n$ assigned to each value.

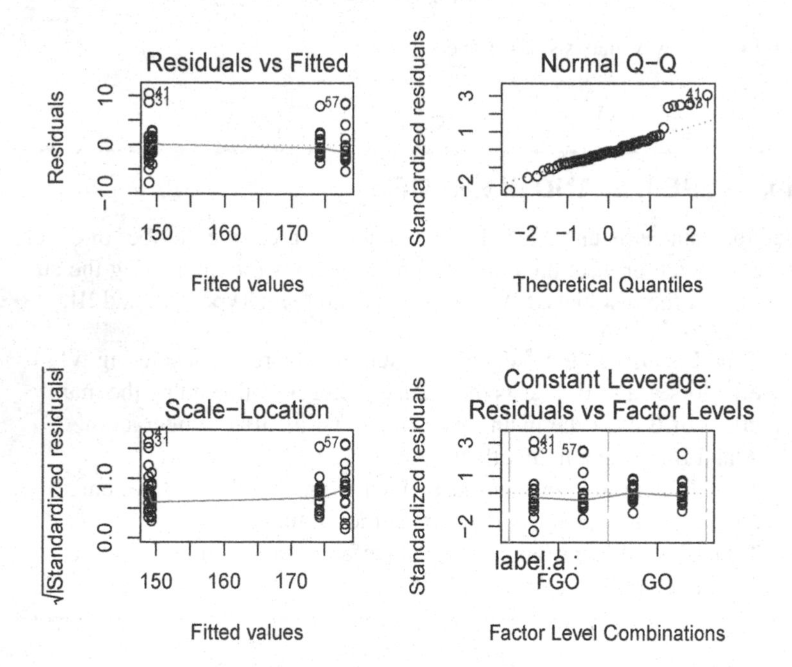

Figure 1.12 Residuals' two way analysis of variance.

1.15.1 TWO-SAMPLE PROBLEMS: COMPARING MEANS OR MEDIAN?

When and how do we compare the median or mean of two sets?
We can make use of the following practical rules:

- Use two sample t-methods so that the data distribution in each set is symmetric and slightly skewed (and the data-set is considered large)

- Compare the medians when the data is not symmetric and/or we have a small data-set, otherwise it is possible to use the means
- If data-sets present extreme values, and they are quite skewed, and the data-set (one or both) cannot be considered large enough, we should recur to an alternative that deals with non-normality

1.16 AN EXAMPLE OF NON-NORMAL DISTRIBUTION

Now that we acquired the basic tools for an exploratory data analysis, we will perform an analysis on the data summarized in table where we reported the summary of 1,000 measurements by two different labs performed to determine the grain dimension of a ceramic compound. The data-set is based on the paper by (12). Our goal will be to determine how the measurements are distributed, and whether we can rule out if they can be considered statistically different or not.

First of all, we report the strip chart and the histogram (since the data gathered is in the region of thousands, for clarity, we did not report the stem plot). It seems that there is no time effect (stripchart), and we can notice that the shape of the histogram looks different in comparison with the previous data-set. The distribution looks "skewed" if we use a normal distribution as a reference. This difference can also be quantified using two descriptors called *skeweness* and *kurtosis*.

Skeweness is self-explanatory, while we can consider an informal definition of kurtosis as the degree of "flatness" in comparison with a normal distribution. Both of them measure deviation from normality, and both the numbers are 0 for a normal distribution.

Table 1.27
Summary of 1000 measurements performed by lab A and B based on the work by (12).

	A	B
Minimum	88.339	105.217
Maximum	638.276	429.709
1. Quartile	199.941	193.051
3. Quartile	294.343	253.161
Mean	253.860	225.954
Median	241.788	220.580
Variance	6078.909	2100.759
Stdev	77.967	45.834
Skewness	0.987	0.713
Kurtosis	1.575	1.011

So far we have answered the first question since we've both visually and numerically determined the shape of the distributions of the two series of measurements.

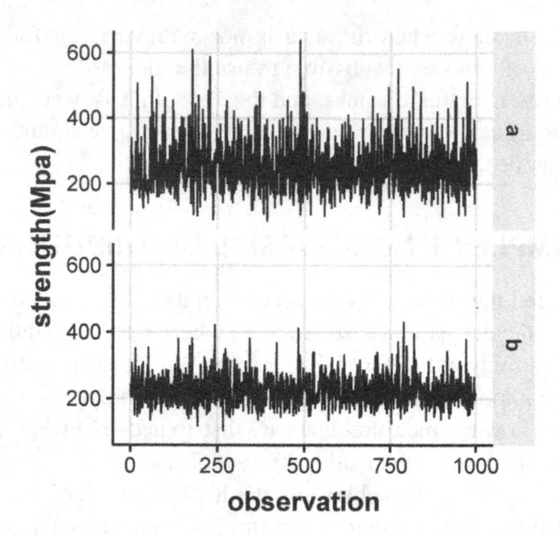

Figure 1.13 Stripchart plot of the dataset based on (12).

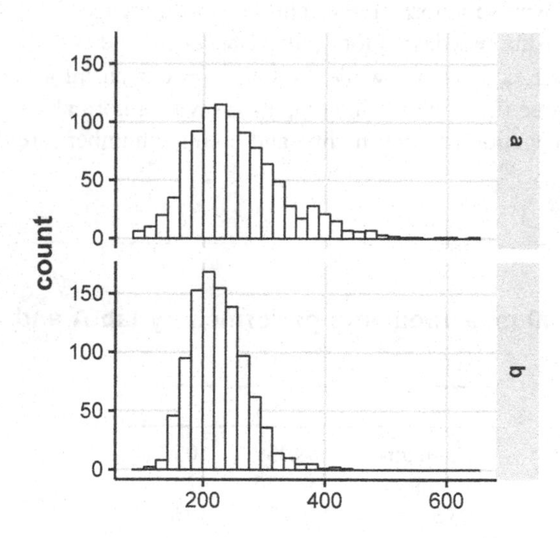

Figure 1.14 Histogram reporting the measurements based on (12).

Now we need to determine if there is a difference among the mean/median of the two series. Can we apply a t-test? Can we use the tools for the analysis of variance explained so far? The answer is no, as you already know. We need to rely on different testing.

Performing the *Kruskaltest* gives us a *p*-value of 0.494, confirming the alternative hypothesis that the measure of location parameters of the distribution of one of the series differs in at least one value comparison from the other series.

1.17 ABOUT VISUAL REPRESENTATION OF DATA

We have presented in this chapter several ways of representing your data. It should be clear at this point that there is no one single technique that can represent all the information stored in a data-set, but that all the tools we have presented help the researcher in highlighting different aspects. In order to provide an intriguing example of this topic, we will present a peculiar data-set provided by Matejka and Fitzmaurice in their paper (9).

All the series have the same value of *1st* quartile, median, *3rd* quartile, equal locations for points 1.5*IQR* from *1st* quartile to *3rd* quartile, and the consequence is that their box plot is the same. This should be a reminder to always start to represent your data with a raw representation (the jitter plot reported) and a histogram.

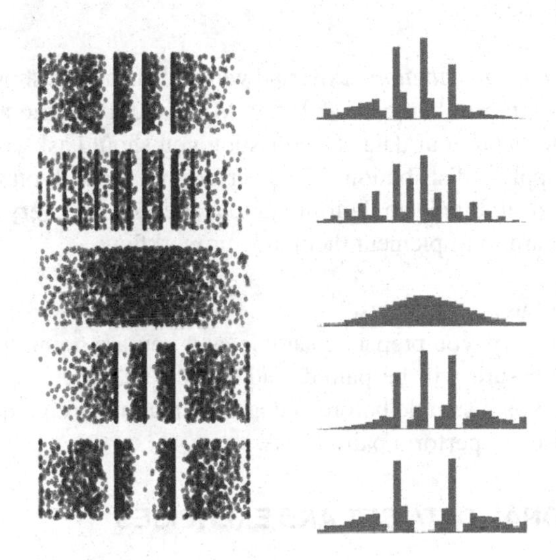

Figure 1.15 Jitter plot and histogram of the data-set.

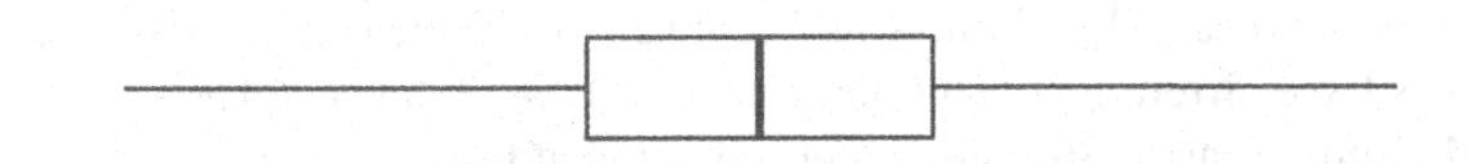

Figure 1.16 Boxplot of data-set.

1.18 FAQ

In this section I will report the FAQ collected during teaching in laboratory and during workshops. Some of them can seem pointless, but it is indeed a good way to repeat concepts that you will need to get accustomed to (*repetita adiuvant*)

Q. Why should I plot my data-set if I can calculate all the descriptors? Or why should I calculate the mean/median/range etc., when I can have a look at the data?

A. Some people prefer visual representation and some of them prefer to have a look at the raw numbers/calculation. A good approach to your exploratory data analysis should include both methodologies. Calculations and plots are now cheap. Can you imagine that scientists used to create representations/ calculations by hand for each plot with paper and pencil? Now in a few seconds, you can represents thousands and thousands of points, so why not?

Q. If I see a trend but the test I applied gave me a p which was not significant, should I reject the hypothesis I made on my data-set?

A. The answer, even if not satisfying, is that performing additional measurements can help in clarifying your doubts, but I would also like to ask if your hypothesis was sound. p is not a magical tool that can solve your problem. Indeed, it can complicate your analysis if misused.

Q. Why do so *many distributions* exist and which distribution should I use?

A. Well, for sure, more than one, or better, you should use the appropriate one while performing tests on your data. The question you should ask yourself is if is it correct to blindly apply a distribution. Since performing a check on the distribution of data is not a big deal, thanks to their ubiquitous presence in every statistical software, you should learn to implement them in your workflow.

Q When should I use a paired test?

A If in the laboratory, you prepare a sample, and you are going to monitor it for some time, your measure will be paired, and so you will use a paired test. Measurement performed on a sample before and after a treatment (e.g., quenching as an example) will be used to perform paired tests.

1.18.1 ADDITIONAL DATA-SET AND EXERCISES

In order to revise the concepts presented in this chapter, the reader is suggested to download the data-set accompanying the book.

The first one is an extended version of data presented in (6). It gathers simulated measurements for 6 samples, from 4 different sets of 50 measurements. The simulated samples have different content percentage of PLA (6 levels). The variables are Young's Modulus, Tensile Strength, and Elongation at break.

The second data-set is based on the work of (3). It is composed by simulated measurements for 8 samples, from 4 different sets of 50 measurements. The simulated samples have different content percentage of a filler (3 levels), and the responses are Pencil Hardness, Scratch Resistance, Hardness (MPA), and Elastic Modulus.

The third data-set is based on (6) and reports 36 observations gathered by 4 different labs on the tensile strength of bottles containing PLA.

The reader is suggested to:

- Plot the data using the techniques shown in the chapter according to samples, replicates, and variables
- Check for the correct number of significant figures for each variables

- Apply a suitable test for checking the normality assumption of each of the variable
- Test for the presence of outliers
- Test the differences of the means/median/mode for each variable according to the different set present in the data-set

1.18.2 REMARKS

As a final remark, in order to perform complete EDA, the readers are suggested always to:

- Create graphs of your data according to the type of variable of your system
- Check the scales of your plot(s). Examine the location of your data. Find the center of the distribution. Examine the shape of your distribution (does it resemble any known distribution?)
- Beware of outliers. While plotting our data it's easy to spot unusual observations or outliers. Keep in mind that outliers are not "errors" to delete from your data-set; in fact, they can record very valuable information. Before blindly ruling them out of your data-set, investigate the cause of their presence
- Make use of appropriate tests to check your assumption on the distributions of the samples and on the ouliers spotted

1.18.3 SUGGESTED ESSENTIAL LITERATURE

In this section I report a minimal essential selection of books that can be useful for the reader.

In order to keep references on basic statistics, I suggest the reader to have a look at *Samuels' "Statistic for the Life Sciences"*. Starting from the concept of *samples* and concluding with *Linear Regression and Correlation* it will present the reader clear explanations of the basic concepts of statistics while offering a wide selection of examples (as the title implies, it is life science-based) (17).

For an approach closest to chemistry, a very useful statistical textbook with an analytical focus is *"Statistics and Chemometrics for Analytical Chemistry"* written by Miller. It's concise, rigorous, and filled with examples. I highly recommend the reader to buy it as a companion for everyday laboratory activities (10).

In order to take advantage of the existing computer tools for performing a statistical analysis and get acquainted with a scientific programming language, I recommend the reader to check the book by Larry Pace *Beginning R, An introduction to Statistical Programming*. It guides the reader from the first steps in the R language explaining its fundamentals and showing examples of how a statistical analysis can be performed (14).

The study of uncertainty can be a daunting task. In order to grab the basics, the reader is suggested to read the *Evaluation of measurements data-Guide to expression of uncertainty in measurements* (2008 edition). Even if not an easy read at first, it is the reference on this topic. Also the book from Radcliffe *Doubt-Free uncertainty in*

Measurement can be a very good introduction for students and practitioners, and can be a valid companion while dealing with this topic.

In order to create plots similar to the one presented in this (and the following) chapters, you can refer to the text from Wickham, "Elegant Graphics for Data Analysis" (19). With its simple approach, it succeeds in explaining the principles behind G-G plot, and will guide the reader with hundreds of recipes.

Bibliography

1. Dennis D. Boos and L.A. Stefanski. *Essential Statistical Inference Theory and Methods*. Springer, 2006.
2. A. C. Davidson and D. Kuonen. An introduction to the bootstrap with application in R. *Statistical Computing & Statistical Graphics Newsletter*, 2003.
3. Hang Hu, Shuyan Zhao, Gang Sun, Yongjian Zhong, and Bo You. Evaluation of scratch resistance of functionalized graphene oxide/polysiloxane nanocomposite coatings. *Progress in Organic Coatings*, 117:118–129, apr 2018.
4. Joint Committee for Guides in Metrology. Evaluation of measurement data — Supplement 1 to the "Guide to the expression of uncertainty in measurement" — Propagation of distributions using a Monte Carlo method. *Evaluation*, 2008.
5. K Krishnamoorthy. *Handbook of Statistical Distributions with Applications*, volume 20064736 of *Statistics: A Series of Textbooks and Monographs*. Chapman and Hall/CRC, jun 2006.
6. F.P. La Mantia, L. Botta, M. Morreale, and R. Scaffaro. Effect of small amounts of poly(lactic acid) on the recycling of poly(ethylene terephthalate) bottles. *Polymer Degradation and Stability*, 97(1):21–24, jan 2012.
7. S.N. Lahiri. *Resampling Methods for Dependent Data*. Springer Nature, 2003.
8. Marcel Maeder and Yorck-Michael Neuhold. *Practical data analysis in chemistry*. Elsevier Science, 2007.
9. Justin Matejka and George Fitzmaurice. Same Stats, Different Graphs. In *dl.acm.org*, pages 1290–1294, 2017.
10. N.Miller Miller and Jane C. Miller. *Statistics and Chemometrics for Analytical Chemistry*. Prentice Hall, 2010.
11. Douglas C Montgomery. *Applied Statistics and Probability for Engineers Third Edition*. John Wiley & Sons, 2003.
12. A Moure, T Hungría, A Castro, and L Pardo. Quantitative microstructural analysis and piezoelectricity of highly dense, submicron-structured NaNbO 3 ceramics from mechanically activated precursors. *Journal of the European Ceramic Society*, 29:2297–2308, 2009.
13. Geoffrey R. Norman and David L. Streiner. *Biostatistics : the bare essentials*. B.C. Decker, 2008.
14. Larry Pace. *Beginning R*. Apress, 2012.
15. Neil J. Salkind. Descriptive Statistics. *Encyclopedia of Research Design*, pages 181–186, 2010.
16. Salkind. J. and Kristin. R. *Encyclopidia of Measurement and Statistics*. Sage Publishing, 2007.
17. Myra L. Samuels, Jeffrey A. Witmer, and Andrew Schaffner. *Statistics for the life sciences*. Pearson, 4th ed., international ed edition, 2012.
18. Iñaki Ucar, Edzer Pebesma, and Arturo Azcorra. Measurement Errors in R. *The R Journal*, 10(2):549, 2019.
19. Hadley. Wickham. *Ggplot2 : elegant graphics for data analysis*. Springer, 2009.

Part II

Essential Multivariate Statistics

2 Design of Experiment

Objectives:

- learn basic concepts of design of experiment
- learn how to select the most suitable design experiment for your investigation
- analyze the data obtained via design of experiment in order to obtain a model of the system investigated

2.1 INTRODUCTION

As presented in (11), when performing an experiment (a single test or series of runs):

> We want to determine which *input variables* are responsible for observed changes in the *response* and then develop a *model* relating the *response* to the input variables and to use this model for process or system development or other decision making.

We can remark that this definition involves that the researcher needs to take the following steps, as also reported by (2)

1. *defining* the *goal* of our study
2. *identifying* the *variables* involved in the process
3. *developing* a *model* of the process
4. *choosing* and *validating* a *design*
5. *analyzing* the results obtained

How can we define the *response* of the system in the exam? We can answer this question also by reporting a quote from (2):

> A well-designed experiment will provide maximum information for a given level of effort. The amount of information provided by the experiment or data collection process can be measured in several ways. Three of those measures are *variance*, *confounding*, and *bias*.

We already presented in the previous chapter definitions of *variance* and *bias*, and in this chapter we will introduce the definition of *confounding*.

An example of design of experiment to reduce variance
How to weigh your samples with a twin-pan balance?
Let's say we need to weigh two samples (A and B). In the first approach we use two sessions, and so we have a resulting variance of σ^2; this is probably the most common approach. Now we use a less common approach. First we weigh them together putting the sample in the same pan ($A+B$), and then we put the weights in opposite pans (and then balance with the needed counter-weight (A-B).
In the latter case, the variance of A is given by var(sum+difference)/2 = $\sigma/2$.
The number of experiments stays the same, but the variance becomes half.
This kind of strategy is used by factorial designs.

2.2 RANDOMIZATION

The first examples that we will report will familiarize the reader with the concept of *randomization*. When we deal with measurements repeated over time, *uncontrolled factors* in the experiment can cause a trend in the response measured, and consequently create a *correlation in the errors*.

This problem can be overcome by introducing *randomization*. One of the classic examples reported by (10) is when we need to measure a quantity as the molarity of a solution.

If we need to plan experiments that involve measuring the solutions affected by the time during the same session of analysis, we should plan the experiment in order to measure different concentrations of different solutions in the same session of analysis, as reported by Table 2.1.

Table 2.1
Experimental plan where the determinations of the values of one of the solutions (0.2 M) are confined to the first two days of analysis.

Day	Sample 1	Sample 2	Sample 3
1	0.2 M	1 M	0.2 M
2	0.2 M	0.2 M	1 M
3	1 M	0.5 M	0.5 M
4	0.5 M	1 M	0.5 M

It is indeed important to notice that the design is still affected by a time effect, since all the measurement with a concentration of 0.2 M will be performed in the first two days of measurement.

A better design is reported in Table 2.2, where each concentration is tested each day, and we do not have all the measurements of one concentration performed over consecutive days.

Table 2.2
Experimental plan where the determinations of the values of all the solutions are randomized through the whole days of analysis.

Day	Sample 1	Sample 2	Sample 3
1	0.2 M	1 M	0.5 M
2	0.2 M	0.5 M	1 M
3	1 M	0.5 M	0.2 M
4	1 M	0.2 M	0.5 M

Each group of concentration will be then considered a *block* (term derived from R.A. Fisher in his agricultural experiment), and so the design adopted and presented here is called *randomized block design*.

The concept of block can be useful also to create other kinds of designs: when there is an *equal* number of blocks and samples (in these cases, days and concentrations) to measure, we can use an experimental design as the one reported in Table 2.3 and in Table 2.4.

Table 2.3
Latin Square Design for three factors.

Day	Sample 1	Sample 2	Sample 3
1	A	B	C
2	C	A	B
3	B	C	A

Table 2.4
Latin Square Design for four factors.

Day	Sample 1	Sample 2	Sample 3	Sample 4
1	A	B	C	D
2	D	A	B	C
3	C	D	A	B
3	B	C	D	A

This design of experiments (and all other generated in the same way) is known as *Latin Square Design*.

The designs that provide measurements for every possible combination of the factors are called *cross classified* designs, while in other cases the designs are called *nested* or *hierarchical*.

2.3 DATA-SET OPT CABLES

In order to introduce the principles of one of the most used kind of design, *factorial design*, we will start showing an application of a simple design of experiment based on the article by (6), where the authors studied the effect of mold temperature and pulling speed on the tensile strength of cylindrical cables produced by pultrusion.

The goal (*first step*) will be to find the conditions that maximize the *tensile strength*:

The *second step* will be to define the factors involved in the study.

The *control factors* are:

- *pulling speed (m/s)* denoted as Factor A
- *temperature* of the mold (°C) denoted as Factor B

and the *response variable* is:

- *tensile strength*

The limits of the control factors are 170-190°C for the temperature of the mold and 0.025 and 0.050 (m/s) for the pulling speed.

From now on we will use the term *low level* (denoted with a "-" or "-1") and *high level* (denoted with a "+" or "+1").

2.4 ONE VARIABLE AT A TIME DESIGN

Before considering an approach based on design of experiment, and so selecting a model (*step three*) and a design (*step four*), we will consider which result we would have obtained using the very common **O**ne **V**ariable **A** **T**ime approach (OVAT or also OFAT, **O**ne **F**actor **A**t **T**ime and even COST, **C**hanging **O**ne **S**igle variable at a **T**ime).

Using an OFAT approach, first of all, we select a starting point, a *baseline* of *levels* for all factors, and then we vary each factor, keeping the other factor constant at baseline levels.

We start with the lower level (170°C), performing one experiment at 0.025 m/s and another one at 0.050 m/s. We obtain a tensile strength of 1130 MPa and 1078 MPa, respectively. Since we obtained the highest impact strength value setting the speed at 0.050 m/s, we fix it and then perform an additional experiment at 190°C, obtaining as result an impact strength of 1398 MPa. We summarize the experimental plan in the following list in order also to highlight the sequential nature of the OVAT process:

1. Measurement at 0.025 m/s with a resulting tensile strength of 1130 MPa
2. Measurement at 0.050 m/s with a resulting tensile strength of 1078 MPa
3. Fix measurement at 0.050 m/s and set temperature to 190°C, obtaining as a response a tensile strength of 1398 Mpa

Please notice that with the experiment performed, we have not *any information* about the *interaction* of the parameters. Also, this was one of the possible paths

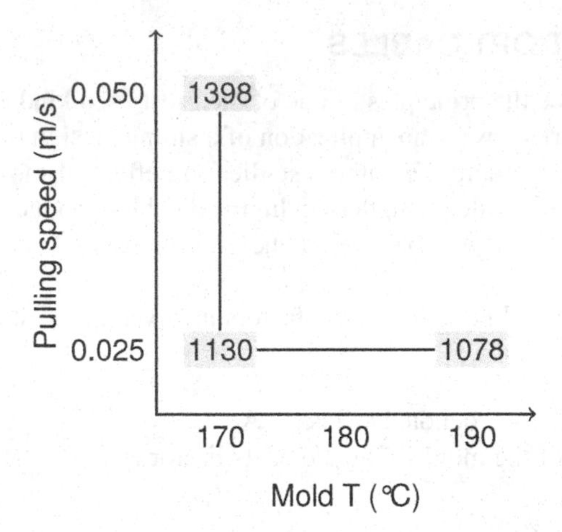

Figure 2.1 Measurement performed using a **O**ne **F**actor **A**t a **T**ime (OVAT) approach based on (6).

since we could have followed an analogous path started by "fixing" the temperature parameters and changing the pull speed.

2.5 FACTORIAL DESIGN

Now we will apply a factorial design 2^2 (reported in Figure 2.2). **Why is it called factorial design?** Factorial designs are designs of experiments *where the factors in the exam are investigated in combinations*. When all possible combinations are present, we have a *complete* factorial design, while in other cases we have *incomplete* factorial designs.

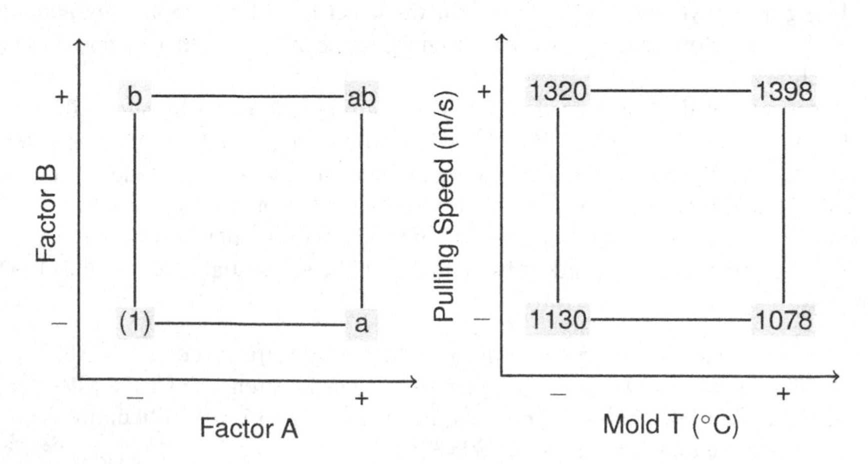

Figure 2.2 Scheme of a 2^2 factorial design and results of the measurements performed on our system.

Table 2.5
Data-set based on the work of Faria et al. (6).

A	B	Pulling Speed (m/s)	Mold Temperature (°C)	Impact Strength (MPa)	rep
-	-	0.0250	170	1130	1
+	-	0.0500	170	1078	1
-	+	0.0250	190	1398	1
+	+	0.0500	190	1320	1

After performing the experiments, we can now quantify the *main effects* for each factor, defined as the *difference between the average response of the low levels—the average response of the higher levels.*

Mold effect (A) =

$$\hat{A} = \frac{a+ab}{2} - \frac{(1)+b}{2} = \frac{1320+1078}{2} - \frac{1398+1130}{2} = 1199 - 1264 = \mathbf{-65}$$

For Pulling Speed effect (B) we have:

$$\hat{B} = \frac{b+ab}{2} - \frac{(1)+b}{2} = 1359 - 1104 = \mathbf{255}$$

where the labels (1), a, b, ab (1), a, b, ab represents the results of the measurement performed according to Figure 2.2. We can notice that the difference in response between the levels of the first factor is not the same for the second one. This suggests that there is an **interaction** between the factors. The magnitude difference can be calculated as the *average* differences of the A effects:

Difference at the low level of B (B^-):

$$a - (1) = 1078 - 1130 = -52$$

Difference at the high level of B (B^+):

$$ab - b = 1320 - 1398 = -78$$

Interaction AB:

$$\hat{AB} = AB = (-78 - (-52))/2 = -13$$

We can also plot these interactions as shown in Figure 2.3. It's important to notice that interaction plots, even if they constitute a valid aid, should not be the *only* technique used to assess the interactions between two factors.

It can also be helpful for the reader to notice that:

- Since the effect of the *pulling speed* is negative, it means that from the low level to the high level, we have an average *decrease* in response of 65 MPa (the higher the pulling speed, the lower the tensile strength)

Table 2.6
Table of Coefficients and the square for the 2 x 2 factorial model in exam.

a	ab	(1)	b	SUM
+1	+1	-1	-1	0
1	1	1	1	4

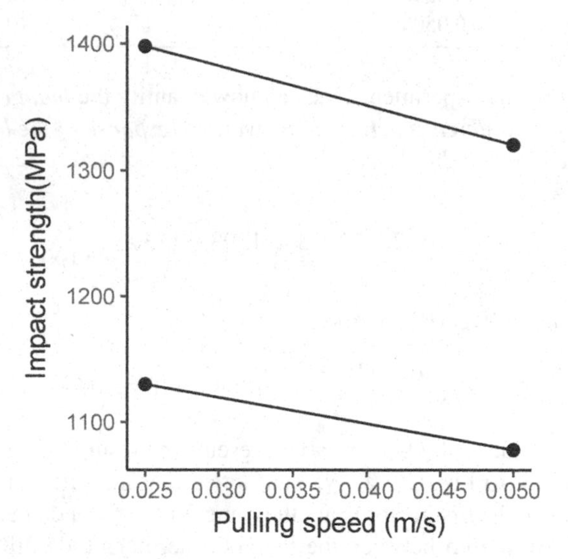

Figure 2.3 Plot of the interactions for the data-set reported in 2.5. The lower line refers to the experiments performed at 170°C, while the upper line refers to experiments performed at 190°C. There is only a weak interaction effect, since the slope of the segments in this part of the experimental domain are similar.

- Increasing the mold temperature from low level to high causes an average response of 255 (the higher the mold temperature, the higher the tensile strength)
- The difference of the average of the response is -13, indicating a difference in the effect of the parameters on the response

2.6 REGRESSION MODEL REPRESENTATIONS

If the design factors chosen are *quantitative*, as in our data-set, another way of studying the interaction among (in our case between) the factors is by means of a **regression model representation**. A regression model of our two-factor factorial experiment can be written as:

$$y = \beta_0 + \beta_1 x_1 + \beta_2 x_2 + \beta_{12} x_1 x_{12} + \varepsilon \tag{2.1}$$

where y is the response, $\beta, \beta_1, \beta_2, \beta_{12}$ are the parameters that we need to determine, x_1 and x_2 are the factors chosen, and x_{12} their interaction. The variables are reported in *coded scale* (-1, +1).

We calculated previously the effects of estimates of parameters A and B. *Regression parameters* are related to them, and can be calculated as:

$$\beta_1 = 255/2 = 127.5$$

and

$$\beta_2 = -65/2$$

The interaction also will be:

$$\beta_{12} = -13/2 = -6.5$$

β_0 is calculated as the average effect of all the responses and so:

$$\beta_0 = (1130 + 1078 + 1398 + 1320)/4 = 1231.5$$

The fitted model will be:

$$\hat{y} = 1231.5 + 127.5x_1 - 32.5x_2 - 6.5x_1x_2$$

Even without applying an ANOVA to the model obtained we can comment that the coefficient of the interaction between factors is lower in comparison with the main effects.

We remind the readers that we created this model using four(!) points. So a laboratory practitioner, before analyzing this model in depth and getting final conclusions, *should increase* the number of experiments performed. We cannot indeed calculate an F value due to the lack of replicates in the design.

It should also be noticed that for now, we did not select a *specific design*. We first chose a probability model for the system or process (selecting variables and making a hypothesis on the relations between the dependent and independent variable), and performed an *exploratory factorial design*.

After this step we will further explore the domain of the experiment, eventually selecting another design.

2.6.1 FACTORIAL MODEL INCLUDING THREE REPLICATES IN THE CENTER

One way to improve the quality of our model will be to add a replicated central point to the experimental domain. In doing so, we assume homoscedascity conditions.

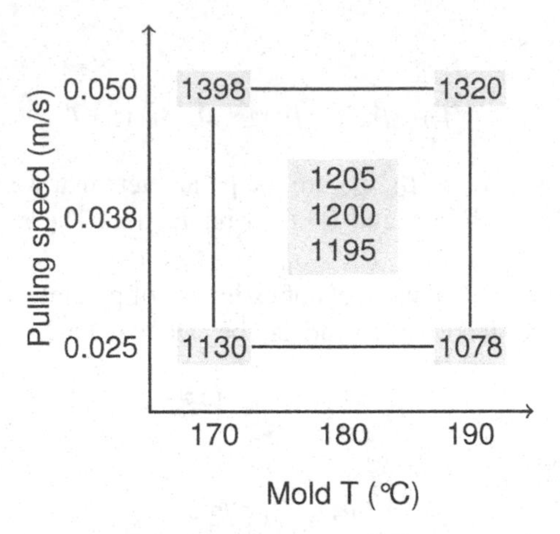

Figure 2.4 Measurements performed according to a 2^3 factorial design, including three replicates in the central point of the design.

> **Homoscedasticity and Heteroscedasticity** It's due to notice that we assume that each sample is from a common population and that samples come from populations with equal variance. We will also call the latter condition (variables with same finite variance, or in other words, homogeneity of variance) *homoscedasticity*, also spelled as *homoskedasticity*, and its complementary *heteroscedasticity* or *heteroskedasticity*.

Analogous to the calculations performed on the data gathered for the 2^2 factorial design, we can calculate the regression models that will include the central points this time, and we can also calculate the *sum of square of the factors*, and perform an Analysis of Variance on the model.

Considering the numerator of the $\hat{A}$, $\hat{B}$, $\hat{AB}$ that multiplied by the sum of the square previously calculated gives us:

$$SS(A) = SS(MoldTemp) = \frac{[\hat{A}]^2}{4} = \frac{255^2}{4} = 16252.25$$

$$SS(B) = SS(PullingSpeed) = \frac{[\hat{A}]^2}{4} = \frac{-65^2}{4} = 1056.25$$

$$SS(AB) = SS(MoldxPullingSpeed) = \frac{[\hat{A}]^2}{4} = \frac{-13^2}{4} = 42.25$$

Multiple R-squared: 0.9754; Adjusted R-squared: 0.9508; F-statistic: 39.65 on 3 and 3 DF; P-value: 0.006503.

In Figure 2.4, we reported a selection of diagnostic plots for the model. The first is a plot of residuals vs fitted plot. It highlights that the model does not fit well point 1 ,4 and 5 due to the high residuals vs the fitted values.

Table 2.7
Data-set based on the work of Faria et al. that includes three replicates in the central point of the experimental domain. A refers to Mold T (°C), and B to pulling speed (m/s).

A	B	Pulling Speed(m/s)	Mold Temperature (C)	Impact Strength(MPa)	rep
-	-	0.0250	170	1130	1
-	+	0.0250	190	1078	1
0	0	0.0375	180	1205	1
0	0	0.0375	180	1200	1
0	0	0.0375	180	1195	1
+	-	0.500	170	1398	1
+	+	0.500	190	1320	1

Table 2.8
Regression Model calculated for data-set based on the work of Faria et al. that includes three replicates in the central point of the experimental domain.

Term	Estimate	Std.error	Statistic	P-value
(Intercept)	1218	9.1313	133.39	9.2905e-07
x1	127.5	12.08	10.555	0.0018165
x2	-32.5	12.08	-2.6905	0.074383
x1:x2	-6.5	12.08	-0.5381	0.62787

In an *ideal Residuals vs fitted plot*, residuals should be randomly and uniformly distributed near the zero line.

The *Normal Q-Q* plot also does highlight trouble with the models since we do not have a distribution on the line highlighted (that represents normality).

The *Scale-Location* plot should be constant in an ideal case, since it is useful for checking variance homogeneity. Again, measurements 1, 6, and 7 are marked as the outliers. As a rule of thumb, we should check samples that show values of 1 of Cook's distance, and deem them suspicious.

Residuals vs Leverage plot. In an ideal case, values of this plot should be scattered around zero.

Cook's distance is the measure of the influence of a single observation on our system and is used to highlight potential outliers. It measures the change of the fitted values when we leave out one observation from the data-set. In this text, values higher or equal than 1 are considered suspicious for practical purposes.

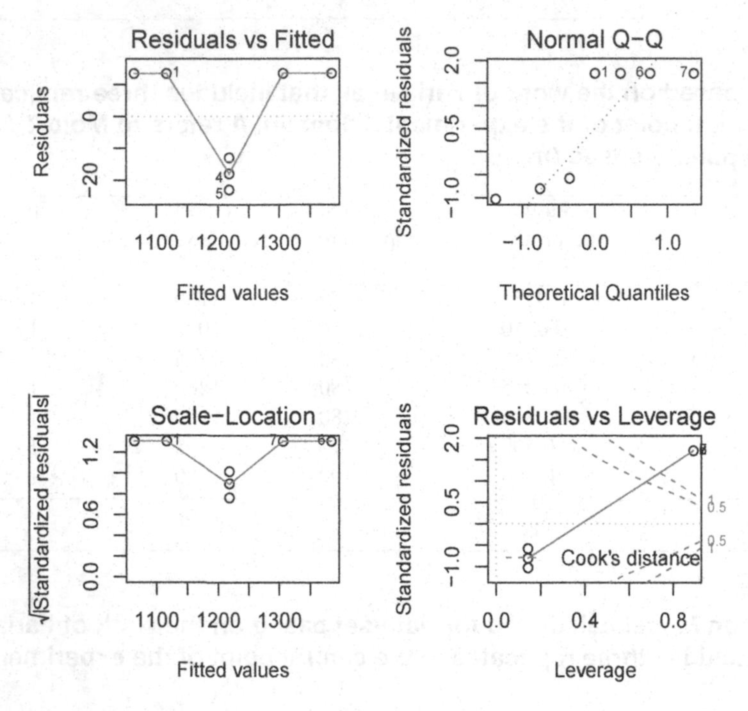

Figure 2.5 Diagnostic plots of the regression model based on the data in 2.7.

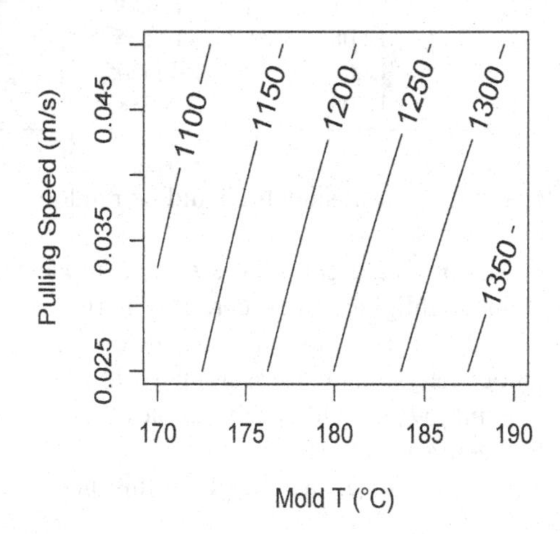

Figure 2.6 2D contour plot and of the regression model based on the data in 2.7.

We can comment that as we can see from the P-value, the interaction among factors is not relevant in comparison with the single effect of each of that factor considered in the study, even if the debate about the use (and abuse) of the parameter p has been rekindled (1).

This model can be also represented using bi-dimensional and three-dimensional plots (**response surface plots**) where we report the factors (in coded or not coded form) and the response. No kind of "distortion" is present in the surface studied confirms a low interaction between factors.

Now that we've performed eight analyses, we should proceed with choosing the best DoE to study the selected answer for the problem. Again we will use a practical approach.

2.6.2 MODEL WITH MORE THAN TWO LEVELS FOR EACH FACTOR

What is the information collected till now?

- Increase of value of the response while increasing both selected factors
- Presence of an interaction between the factors which is lower in comparison with each main effect contribution

The diagnostic plot hints at a few suspect measurements, but considering the number of the samples studied, it would be unwise to leave measurements out of the model.

What are the values of the diagnostics for the model? Multiple R-squared (0.9754), Adjusted R-squared (0.9508), F-statistic: 39.65 on 3 and 3 DF, p-value: 0.006503 represent a very good fit, but due to the number of experiments performed, it would be over optimistic to consider this a "good" model.

One (very simple) way to enhance the quality of the model obtained would be to increase the number of experiments performed.

We will present a regression model for the data-set considering replicates (three) performed at more than two levels for each factor. This model will also let us study the interaction between parameters and highlight the (eventual) presence of quadratic effects due to the presence of replicates.

Looking at Table 2.9 and Table 2.6.2 we can observe that:

- The interactions and the square of the second factor are not significant in comparison with the main effects and the square of the second factor
- Residuals vs Fitted plot reports high values of residual
- The normality still suffers from a bias and the Cook's plot looks similar to the previous one
- We did not report residual vs leverage since the leverage for all factors is the same and is equal to 0.11 (the DoE is *balanced* and the effect of leaving out each one of the measurements has the same effect on the model)
- A curvature in the response surface is now visible due to the higher resolution used for creating the model
- The surface obtained does not present any stationary point, and consequently the minimum and maximum points are situated at its boundaries

The increased number of measurements enhanced for sure the quality of the model obtained (we have more degrees of freedom, and consequently the variance of the system was studied more in depth), but the most important observation is that increasing the number of levels studied highlighted that we indeed have a quadratic

Table 2.9
Extended data-set based on the work of Faria et al. that includes three replicates for each point explored of the experimental domain. In order to avoid systematic errors, the experiments are performed following a random order. For the sake of clarity, without randomization.

A	B	Pulling Speed (m/s)	Mold T(°C)	Impact Strength (MPa)	Rep
-	-	0.025	170	1130	1
-	-	0.025	170	1136	2
-	-	0.025	170	1140	3
0	-	0.0375	180	1100	1
0	-	0.0375	180	1106	2
0	-	0.0375	180	1102	3
+	-	0.05	190	1078	1
+	-	0.05	190	1086	2
+	-	0.05	190	1080	3
-	0	0.025	170	1230	1
-	0	0.025	170	1240	2
-	0	0.025	170	1220	3
0	0	0.0375	180	1205	1
0	0	0.0375	180	1200	2
0	0	0.0375	180	1195	3
+	0	0.05	190	1125	1
+	0	0.05	190	1130	2
+	0	0.05	190	1135	3
-	+	0.025	170	1390	1
-	+	0.025	170	1398	2
-	+	0.025	170	1395	3
0	+	0.0375	180	1340	1
0	+	0.0375	180	1342	2
0	+	0.0375	180	1350	3
+	+	0.05	190	1320	1
+	+	0.05	190	1330	2
+	+	0.05	190	1335	3

effect, which seems more important than the interaction effects between the factors. Even if the number of experiments to perform is still "manageable", we can comment that since we considered a framework of *homoscedasticity*, it was not necessary to eventually perform in triplicate each measurement, and that a reduced number of experiments can give us the *same* practical results.

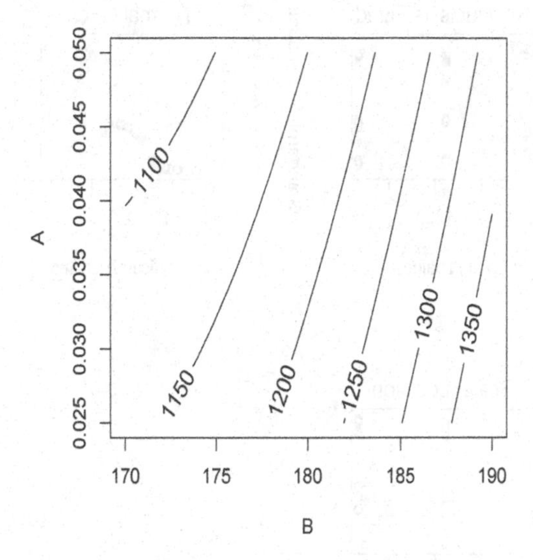

Figure 2.7 2D contour plot of the regression model based on the data in 2.7. A refers to concentration and B to temperature.

Term	Estimate	Std.error	Statistic	P-value
(Intercept)	1186	5.6877	208.52	2.5984e-36
x1	124.56	3.1153	39.982	2.6473e-21
x2	-36.667	3.1153	-11.77	1.0396e-10
x1:x2	-3	3.8154	-0.78629	0.44048
x1^2	44.333	5.3958	8.2163	5.3556e-08
x2^2	1	5.3958	0.18533	0.85475

2.7 DATA-SET EMAGMA, AN EXAMPLE OF DOE WITH THREE FACTORS

In order to better clarify the concepts of factorial design, we will extend its use to a data-set based on the paper from (12), where the authors studied the condition of a reactive extrusion route to compatibilize blends of poly trimethylene terephthalate (PTT) and polylactic acid (PLA) with the addition of a random terpolymer of ethylene, methyl acrylate, and glycidyl methacrylate (EMAGMA).

The complete data set is reported in Table 2.10.

As we've done in the previous example, we follow the step we've stated while dealing with a DoE workflow.

The first goal (*first step*) will be to find the conditions that maximize the *impact strength*.

The *second step* will be to define are the factors involved in the study.

The *control factors* are:

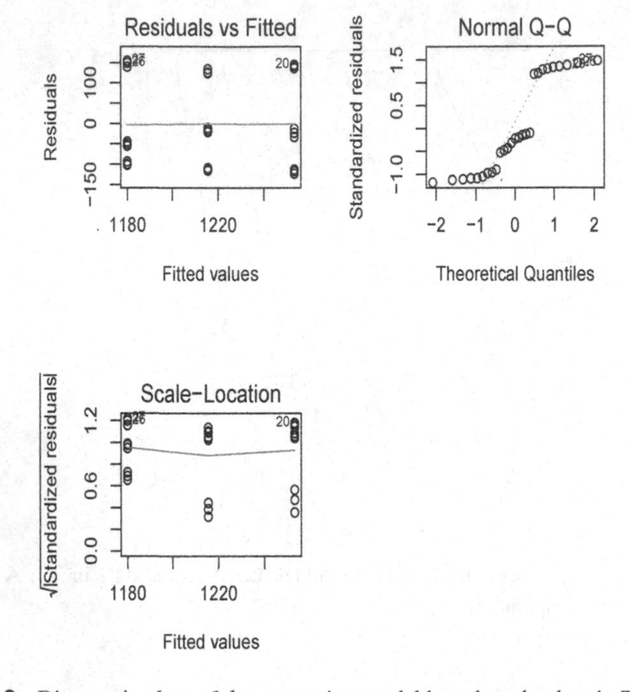

Figure 2.8 Diagnostic plots of the regression model based on the data in Table 2.9.

- *concentration* of terpolymer EMAGMA
- *part per hundred of chain extender* CE
- *screw speed*

and the *response variables* are:

- *impact strength*
- *tensile strength*

2.7.1 WORKFLOW USING OVAT

Again we compare a univariate approach vs a multivariate one. Starting with an OVAT design, one of the paths to follow could be:

- Fix the concentration of CE and screw speed to 0% and 100 rpm, respectively
- Perform the experiment at concentration of terpolymer 10 and 20 wt%, obtaining an impact strength, respectively of 34 and 58 (MPa)
- Fix the EMAGMA% at 20 wt%, since he gave the higher impact strength
- Fix the screw speed to 100 rpm, obtaining an impact strength of around 58 and 125 MPa
- Change the rpm to 200 rmp, obtaining an impact strength of **200 MPa**

We conclude that the best conditions to increase the impact strength is conducted using **20 wt%** with **0.50 phr** of CE and a screw speed of **200 rpm**.

An *alternative path* could be:

- Fix the concentration of CE and the concentration of terpolymer to 0% and 10 wt%
- Perform the experiment at a speed of 100 rpm and 200 rpm, obtaining an impact strength, respectively of 34 and 52 (MPa)
- Fix the speed to 200 rpm (since it gave higher results) and change the concentration of CE, obtaining 52 and 70 MPa
- Finally change EMAGMA% and test 20% (since the experiment at 10% was already performed), obtaining **200 Mpa** as the final value

In this *specific case* we obtain the same value following two different OVAT paths. This is not always the case. In Figure 2.9, we reported the two paths followed.

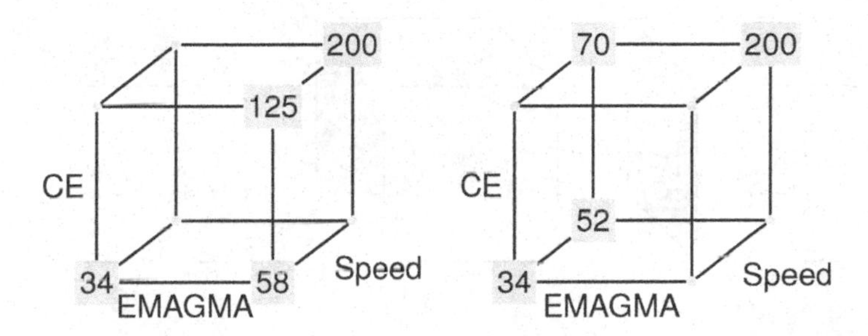

Figure 2.9 Results obtained performing two different OVAT designs. In this specific case, we would obtain the same results.

2.7.2 FACTORIAL DESIGN 2^3

In Table 2.10 and Figure 2.10, we report the experimental results and plots obtained by applying a 2^3 full factorial design.

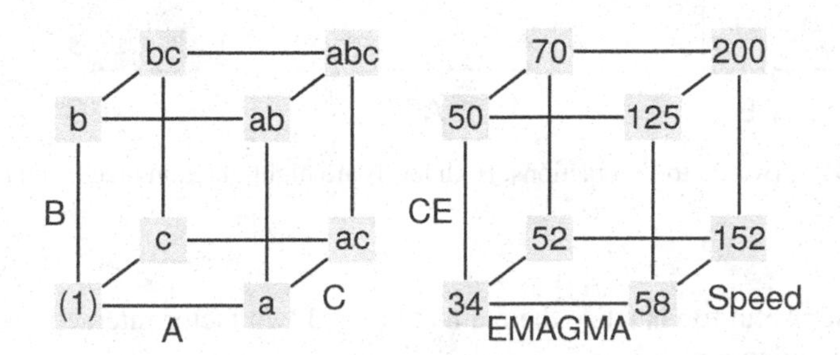

Figure 2.10 Representation of the experiment performed in 2.10 based on the work of Nagarajan et al. (12).

Table 2.10
Data-set based on the work of Nagarajan et al. (12).

A	B	C	EMAGMA (%wt)	CE (phr)	Screw Speed (rpm)	Impact Strength (MPa)
-	-	-	10	0	100	34
-	-	+	10	0	200	52
-	+	-	10	0.5	100	50
-	+	+	10	0.5	200	70
+	-	-	20	0	100	58
+	-	+	20	0	200	152
+	+	-	20	0.5	100	125
+	+	+	20	0.5	200	200

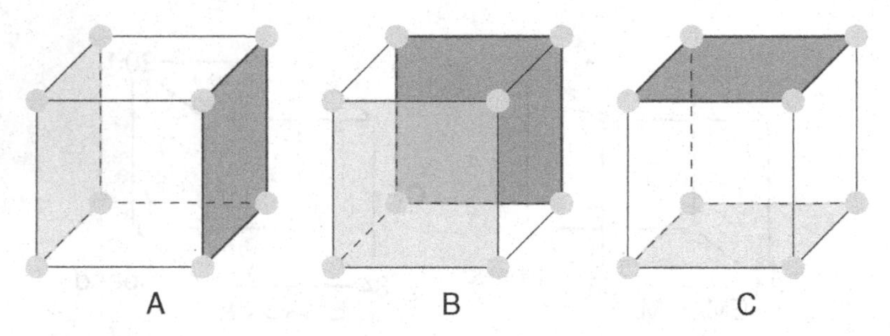

Figure 2.11 Main effects and high levels are highlighted in black; low levels in light gray. A, B, and C refer to the factors in the exam.

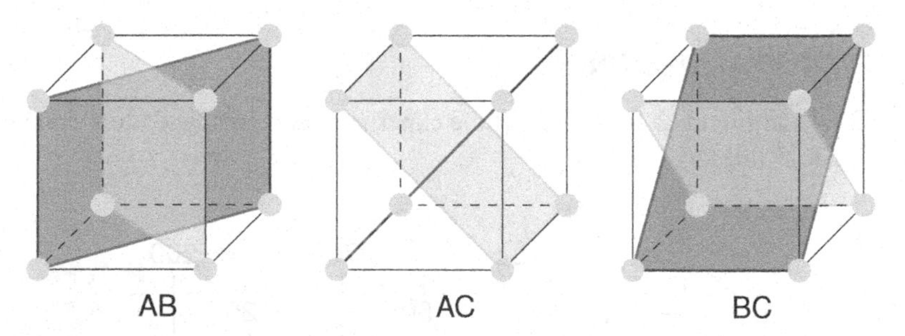

Figure 2.12 Two-factor interactions. High levels are highlighted in black, and low levels in light gray.

It is also useful to visualize the main effect and two factor interactions in a cubic experimental domain.

Low (in light gray) and high levels (darker gray) of the factors will be represented by planes as well as their interactions.

- *concentration* of terpolymer EMAGMA (5.15 wt%)
- *part per hundred of chain extender* CE (0,0.5 phr)
- *screw speed* (100,200 rpm) screw speed

Applying a multivariate regression, we obtain the first exploratory model, giving the following results.

Table 2.11
Model based on data reported in 2.11 for the data-set based on Nagarajan et al. (12).

Term	Estimate	Std.error	Statistic	P-value
(Intercept)	92.625	2.625	35.286	0.018037
x1	41.125	2.625	15.667	0.04058
x2	18.625	2.625	7.0952	0.089138
x3	25.875	2.625	9.8571	0.064364
x1:x2	10.125	2.625	3.8571	0.16149
x1:x3	16.375	2.625	6.2381	0.10119
x2:x3	-2.125	2.625	-0.80952	0.56677

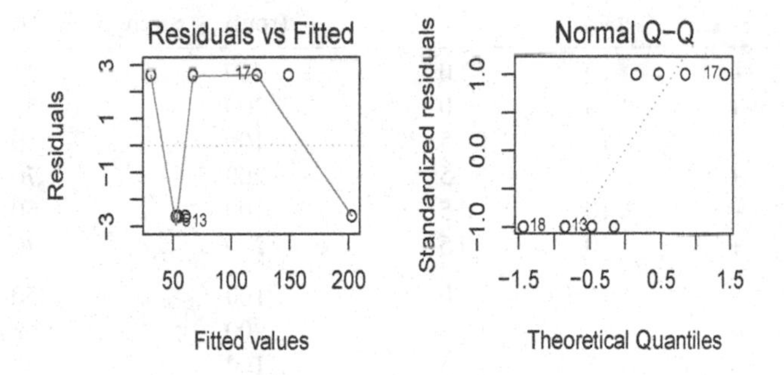

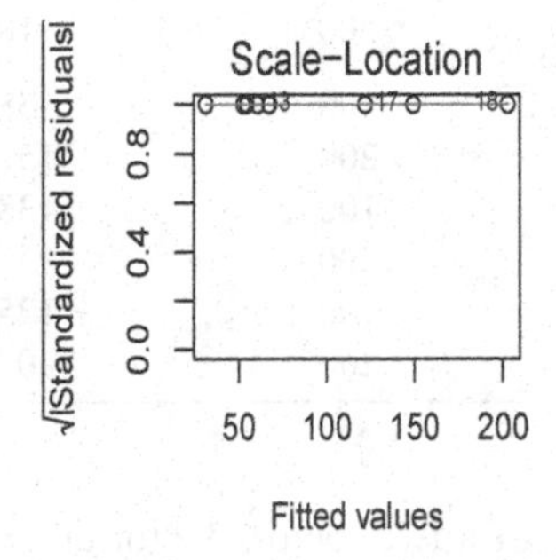

Figure 2.13 Diagnostic plots of the regression model based on the data in 2.11.

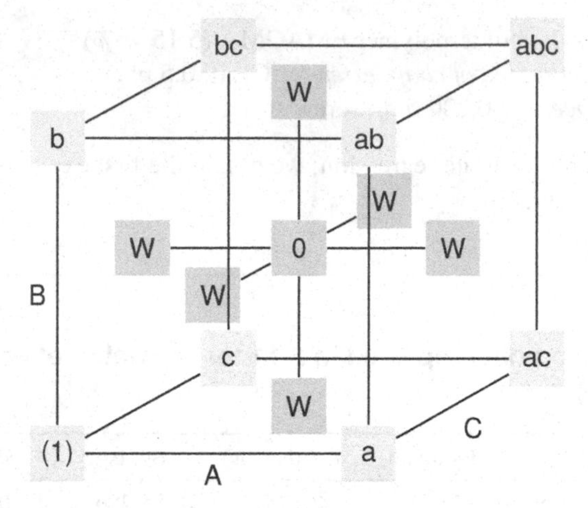

Figure 2.14 Factorial design including center points.

Table 2.12
Data-set based on the work of Nagarajan et al. (12).

A	B	C	EMAGMA (wt%)	CE (phr)	Screw Speed (rpm)	Impact Strength (MPa)
-	-	-	10	0	100	34
-	-	+	10	0	200	52
-	0	-	10	0.25	100	40
-	0	+	10	0.25	200	70
-	+	-	10	0.5	100	50
-	+	+	10	0.5	200	70
0	-	-	15	0	100	50
0	-	+	15	0	200	68
0	0	-	15	0.25	100	92
0	0	+	15	0.25	200	100
0	+	-	15	0.5	100	93
0	+	+	15	0.5	200	118
+	-	-	20	0	100	58
+	-	+	20	0	200	152
+	0	-	20	0.25	100	138
+	0	+	20	0.25	200	170
+	+	-	20	0.5	100	125
+	+	+	20	0.5	200	200

Analogous to what we've done in the case of a two factors factorial design, we will perform an analysis in the *experimental domain center*, and also at the *center of the domain of each pair of factors* in order to study the curvature of the surface obtained (see 2.14 and 2.12).

Table 2.13
Data-set based on the work of Nagarajan et al. (12).

Term	Estimate	Std.error	Statistic	P-value
(Intercept)	93.556	3.936	23.769	8.3254e-11
x1	43.917	4.8206	9.1101	1.8607e-06
x2	20.167	4.8206	4.1834	0.0015278
x3	18	3.936	4.5731	0.00079927
x1:x2	10.125	5.904	1.7149	0.11436
x1:x3	11.083	4.8206	2.2991	0.042095
x2:x3	-0.83333	4.8206	-0.17287	0.86589

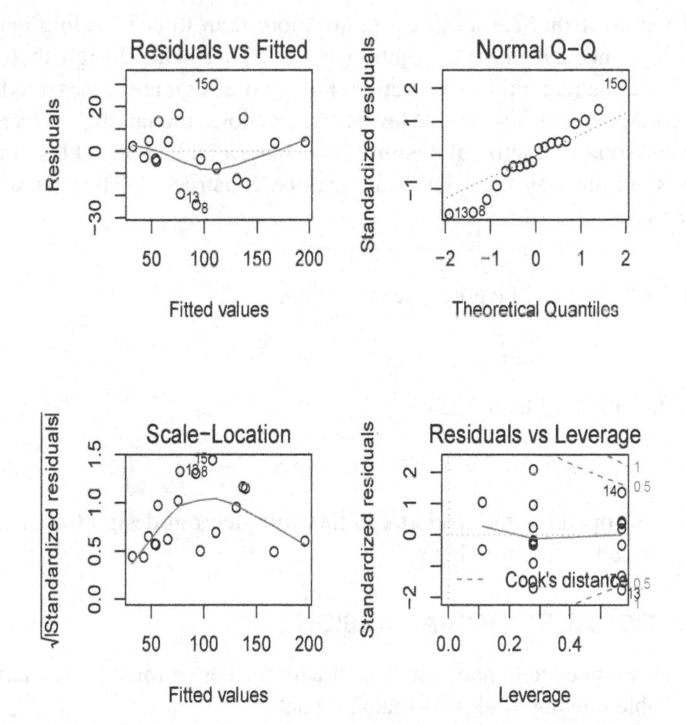

Figure 2.15 Diagnostic plots of the regression model based on the data in 2.12.

The results of the computed model are reported in 2.13 and the diagnostic plots in 2.13. The extra (in comparison with the previous) points in the model make it more reliable in comparison with the previous one, even if can say that:

- The residuals are still high (compare the previous scale values)
- Leverage vs Standardized residuals still indicate suspicious values (3 and 14)

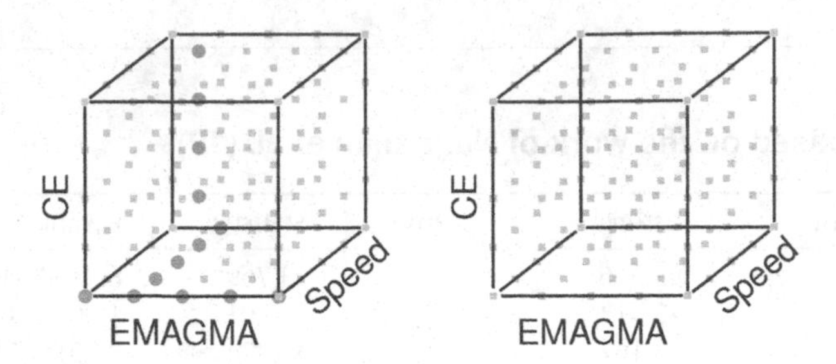

Figure 2.16 Comparison 2.

2.7.3 FACTORIAL DESIGN 2^K

What should we do if the factors in exam are more than three? Adding one more factor would have needed the investigation via a 2^4 **factorial design** that can be represented by a cube (actually a hypercube) Factorial design are represented by the form s^k where s denotes the number of levels and k denotes the number of factors. So **How can we construct a factorial design?** If we have a look at the tables, it is quite easy to understand the way factorial design can be constructed. The first sequence for the first factor is:

$$- + - + - + - + \ldots$$

the sequence for the second factor is the following:

$$- - + + - - + + - - + + \ldots$$

The sequence for the third factor is:

$$- - - - + + + + - - - - + + + + \ldots$$

and so on. In the Appendix, the readers can find full factorial design for five factors and also the R code for creating them.

2.7.4 FRACTIONAL FACTORIAL DESIGN 2^{K-1}

What happens if we need to expand the design with further factors? Is it a viable path to study our problem using a full factorial approach?

As the number of factors increases, the number of the experiments to be performed for the design also increases (i.e., for a complete design of 2^6 we require 64 runs without considering any replicate), so more parsimonious designs, requiring less experiments can be quite useful when dealing with more than 4 factors.

How can we achieve this? A solution is to rely on a *fraction* of factorial design. In a 2^3 factorial design that requires 8 experiments, it is quite easy. Since we can represent the experimental domain with a cube, we can select just half of the vertices, and so run 4 experiments (or 2^{3-1} 2^2 design), selecting one of the two possible fractions ($n = 2$ order) that represents *each factor twice* at its *high level* and *low level*. In the case of four levels the experimental design would be since from the multivariate regression previously reported we know that we can calculate the main effect for each factor easily.

Table 2.14
Fractional factorial design, 2^{4-1}.

A	B	C	D
+	+	-	-
-	-	-	-
-	+	+	-
+	+	+	+
-	+	-	+
+	-	-	+
+	-	+	-
-	-	+	+

- *F*actor A is given by $y_1 + y_2 + y_3 + y_4 - y_5 - y_6 - y_7 - y_8$
- *F*actor B is given by $y_1 + y_2 - y_3 - y_4 + y_5 + y_6 - y_7 - y_8$
- *F*actor C is given by $y_1 - y_2 + y_3 - y_4 + y_5 - y_6 + y_7 - y_8$
- *F*actor D is given by $y_1 - y_2 - y_3 + y_4 - y_5 + y_6 + y_7 - y_8$

We can observe that if we calculate the interaction among AB and D (called ABD for short) obtained by multiplying the signs in the column A, B, and D, we obtain the same signs of column C, or in other words, that the code for ABD is the same as C. This means that C and ABD are *aliases* of each other and that there is a *confounding* among them. We can also calculate that each of the factors in the exam is confounded with the three-fold interactions of the other three factors, as in ALL 2^{k-1} designs. So we spared on the number of experiments to perform, but as a trade off we cannot *resolve* the confounding among the factors. The extent of the confound is reported in roman numbers (**R**) . Considering that the main effects have $p = 1$, second interaction effects have $p = 2$, three-fold have $p = 3$, etc., we have no confounding between a chosen factor p and an effect that contains $< (R - p)$ factors. When we have as an example a fractional factorial of resolution IV, we will have no confounding for p1 and IV(4)-1, with less than three-fold interactions.

2.7.5 ON GRAPHICAL REPRESENTATION OF FACTORIALS WITH FOUR FACTORS

An interesting visual approach to learning how factorial designs work is reported by (2). We just report a brief introduction in order to encourage the reader to explore more of this topic, which is not widely present in literature. To represent a factorial with four factors, graphical representation of lower factorial design can be combined in hierarchical approach.

As an example, combining a one-factor two-levels design with a three-factor two levels design (represented by a cube) can give origin to a 2^4 design, as represented in 2.17.

In this case, the first level of hierarchy (cube) represents the first three factors, while the location of the short line segment represents the factor D. We remind the reader that this is one alternative to represent designs with more than three variables, but that another solution would have been to split the hierarchy by grouping factors in two experiments. By convention, the base level of the hierarchy is called *first level*, or in the presence only two levels, is called *inner design*. It is interesting to notice that design-plots are effective for factorial design, while *D-optimal* design for six factors with more than three levels or factors could not be viewed with these tools. See the end of this chapter for the definition of *optimal design*.

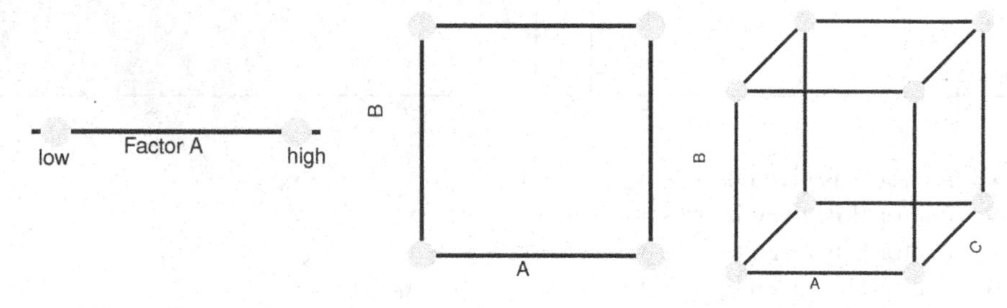

Figure 2.17 Representation of one factor, two factors, and three factors at two levels.

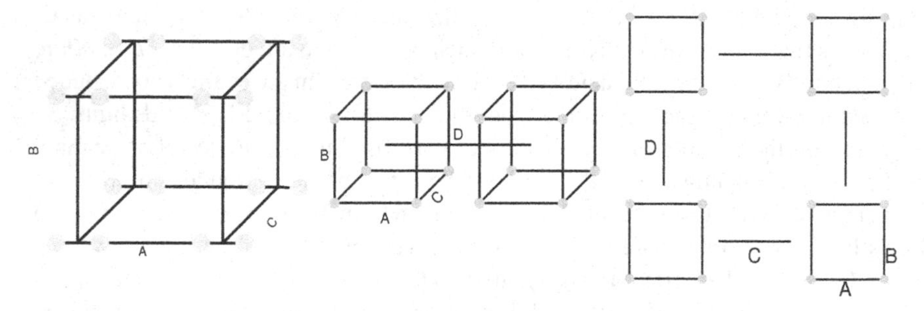

Figure 2.18 Example of design-plot for two levels of factorial design.

2.8 MIXTURE DESIGN

It is very common while working on material synthesis and characterization to deal with mixtures. If we think the components of the mixture as variables with limits depending on each other, we can successfully apply the principles already described while presenting factorial designs. Again with the help of an example, we will show how the reader can approach these kinds of problems.

2.8.1 DATA SET HIPS

We will use a data-set based on the work by (16), where the authors report the study of influence of weight percentage of high impact polystyrene (HIPS), nanosilica, and hardener content on damping mode properties of $epoxy/HIPS/SiO_2$ hybrid nanocomposite.

The data-set is reported in Table 2.15.

Table 2.15
Dataset from (16), where x_1 represents HIPS, $wt\%$ x_2 the concentration of ($SiO_2\%$), and x_3 the hardener content). The concentration is expressed as a fraction of unit, i.e., 1.00 means 100%.

HIPS Content(x_1)	SiO2 Content (x_2)	Hardener Content (x_3)	Damping 1st	Damping 2nd
0.000	1.000	0.000	2.12	3.14
1.000	0.000	0.000	1.71	1.02
0.000	0.000	1.000	3.46	3.64
0.167	0.667	0.167	3.05	3.96
0.500	0.000	0.500	3.41	3.23
0.000	0.500	0.500	3.72	4.61
0.333	0.333	0.333	3.21	4.23
0.167	0.167	0.667	3.74	4.48
0.500	0.500	0.000	2.48	3.51
0.667	0.167	0.167	2.35	2.62

We need to remember that the sum of the three fractions involved has to be one and so the experimental valid points are located on a triangular space across (0,0,1), (0,1,0), and (1,0,0) According to the figure presented, each head and its opposite side indicate the highest and the lowest values of a fraction in the formulation, respectively. After testing the samples, responses are represented in this triangular space as contours where each one of them represents a response obtained with different formulations.

After coding the variables and recording the response, we can calculate a *linear model* for each one of the properties investigated. The model reported in this example takes the name of *three component, simplex-centroid mixture design*.

We reported the regression coefficients for two interested responses in 2.16 and 2.17. We can notice that considering as a limit a P-value $> 95\%$ for the first response (damping 1), the only significative terms are $x1$, $x2$, $x3$, the interaction between $x1$ and $x3$, and between $x2$ and $x3$, while for the second response (damping 2), the significative factors are $x1$, $x2$, $x3$, the interaction between $x1$ and $x2$, $x2$ and $x3$, and finally between $x2$ and $x3$. From 2.19, we can notice that the increasing the portion of $x1$ (HIPS) had a reverse effect on damping, and that a lower value of this variable would have a better effect on the response studied. An increasing value of

Table 2.16
Model obtained for response Damping 1st.

Term	Estimate	Std.error	Statistic	P-value
x1	1.6293	0.17931	9.0869	0.0028158
x2	2.1646	0.17931	12.072	0.0012232
x3	3.4864	0.17931	19.444	0.00029716
x1:x2	2.1874	0.90243	2.4239	0.09384
x1:x3	3.1908	0.90243	3.5358	0.03848
x2:x3	3.8629	0.90243	4.2805	0.023422
x1:x2:x3	-7.2705	5.9537	-1.2212	0.30923

Table 2.17
Model obtained for response Damping 2nd.

Term	Estimate	Std.error	Statistic	P-value
x1	0.93651	0.24642	3.8004	0.031996
x2	3.1037	0.24642	12.595	0.0010793
x3	3.7254	0.24642	15.118	0.00062833
x1:x2	5.4788	1.2402	4.4175	0.021533
x1:x3	3.6038	1.2402	2.9057	0.062212
x2:x3	4.9786	1.2402	4.0143	0.027748
x1:x2:x3	-0.53957	8.1822	-0.065944	0.95157

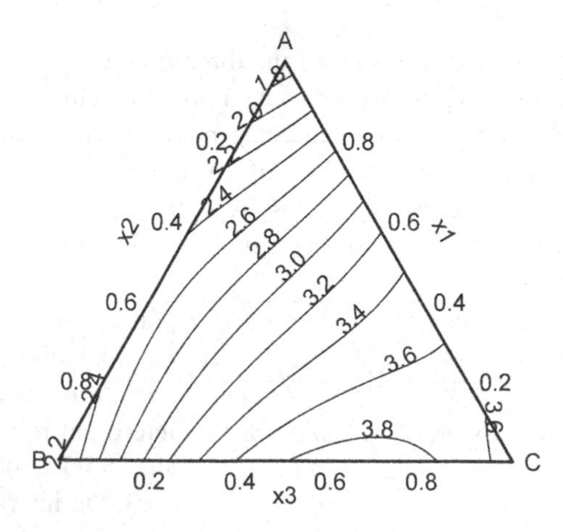

Figure 2.19 Model obtained for 1st damping (16).

$x2$ increases the damping properties to a maximum. Also, from 2.20, we can confirm a reverse effect of the $X1$ values of damping.

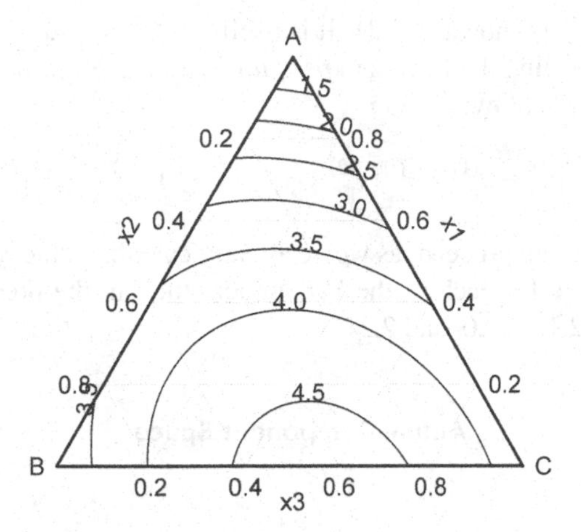

Figure 2.20 Model obtained for 2nd damping (16).

2.9 DESIGN OF EXPERIMENTS MATRIX VS REAL EXPERIMENTS PERFORMED

It is due to notice that the practitioner in the calculation of the model **must** use the **real** concentration employed, i.e., if in theory I needed to use a mixture of $0:33\,x1+0:33\,x2+0:33$, but for some reason, the concentrations of my final solutions are $0:36\,x1+0:42\,x2+0:35$, I **must** use the latter ones in the model(s).

2.9.1 MIXTURE DESIGN IN CONSTRAINED REGION

On the basis of previous experiments or according to the system in exam, it is very common to work on a restricted factor space of chosen composition.

2.9.2 DATA SET CPCB

In order to report an example of this case we will refer to (5). The authors studied the possibility of the use of cement bonded particle board (CPCB) as an additive in sand-cement composites. The factors of the mixture in exam were the concentration of concrete (z1), sand (z2), and CBPB (z3), while the responses the 28-day long strength in centric compresson R_c (Y_1), the bending R_b (Y_2), the flow of water for mixing of dry mix of all components to the state of the equal flowability w in kg for 1 kg of dry mix of all components (Y_3), the density ρ ($kg \cdot m^{-3}$)(Y_4). On the basis of previously performed experiments, the authors selected a sub-area of the design space characterized by the following vertices:

1. A_1 ($x_1 = 0.333, x_2 = 0.333, x_3 = 0.333$)
2. A_2 ($x_1 = 0.250, x_2 = 0.750, x_3 = 0.000$)
3. A_3 ($x_1 = 0.143, x_2 = 0.000, x_3 = 0.8570$)

The selected area is shown in 2.21. It is useful to transform it to a complete simplex design transforming the factors in *pseudofactor* z_1 , z_2, z_3 that are related to the natural factors $x1$, $x2$, $x3$ by:

$$x_i^{(i)} = x_i^{(1)} + z_2^{(i)}(x_i^{(2)} - x_i^{(1)}) + z_3^{(i)}(x_i^{(3)} - x_i^{(1)})$$

At this point, we can proceed, as we've already calculated the regressions models and surface response for each of the $Y(i)$ under study in the previous example, as presented in 2.18, 2.19, 2.20, and 2.21.

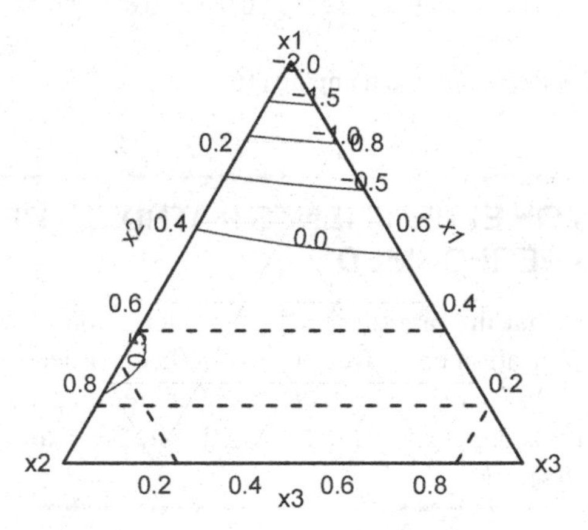

Figure 2.21 Dashed lines represent the limits for the factors in exam reported by (5).

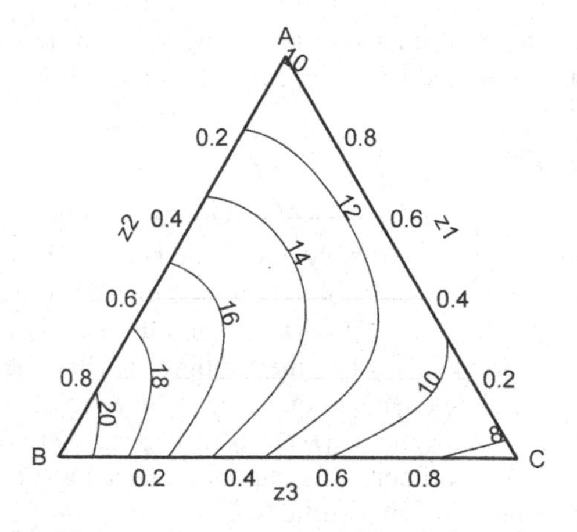

Figure 2.22 Strength of the centric compression of $Y_1 concretesamples$, depending on the content of the components expressed by the pseudofactors- concentration of cement (z_1), sand (z_2), and addition of CBPB waste products (z_3) (5).

Table 2.18
Model obtained for response Y_1.

Term	Estimate	Std.error	Statistic	P-value
z1	9.8412	1.631	6.0338	0.10456
z2	22.083	1.6312	13.537	0.046941
z3	7.5191	1.6302	4.6124	0.13592
z1:z2	-0.39845	7.9822	-0.049918	0.96825
z1:z3	8.4487	7.9903	1.0574	0.48226
z2:z3	-14.055	7.9116	-1.7764	0.3264
z1:z2:z3	54.276	49.92	1.0872	0.4734

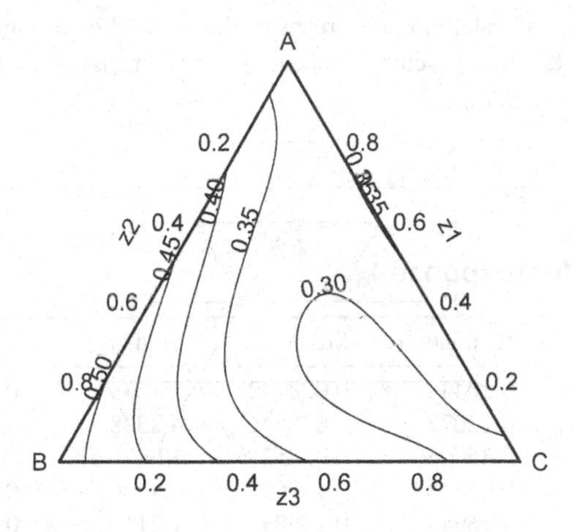

Figure 2.23 Bending strength of Y_2 concrete samples, depending on the content of the components expressed by the *pseudofactors*- concentration of cement (z_1), sand (z_2), addition of CBPB waste products (z_3).

Table 2.19
Model obtained for response Y_2.

Term	Estimate	Std.error	Statistic	P-value
z1	0.32593	0.0487	6.6926	0.094425
z2	0.52339	0.048708	10.745	0.059075
z3	0.28721	0.048677	5.9004	0.10688
z1:z2	0.10763	0.23834	0.45158	0.72997
z1:z3	0.17638	0.23859	0.73927	0.59473
z2:z3	-0.1817	0.23624	-0.76914	0.58261
z1:z2:z3	-2.3221	1.4906	-1.5578	0.3633

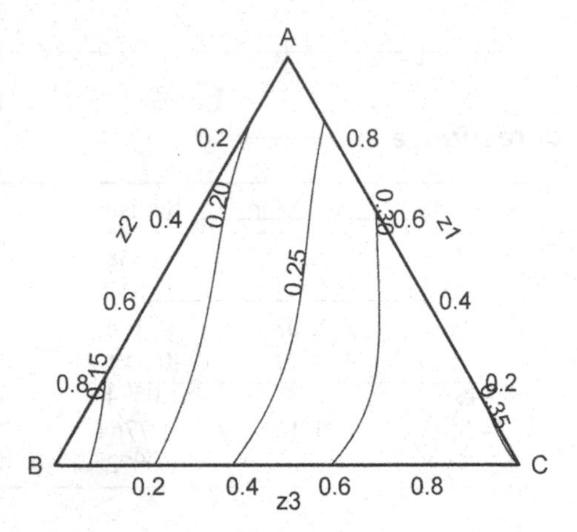

Figure 2.24 Amount of water in Y_3 concrete samples, depending on the content of the components expressed by the pseudofactors- concentration of cement (z_1), sand (z_2), and addition of CBPB waste products (z_3).

Table 2.20
Model obtained for response Y_3.

Term	Estimate	Std.error	Statistic	P-value
z1	0.20445	0.028542	7.1633	0.088302
z2	0.12077	0.028546	4.2308	0.14776
z3	0.35005	0.028528	12.27	0.051768
z1:z2	0.065183	0.13969	0.46664	0.72205
z1:z3	0.16943	0.13983	1.2117	0.43924
z2:z3	0.17975	0.13845	1.2983	0.41783
z1:z2:z3	-0.66851	0.87359	-0.76524	0.58417

We obtain the following equations:

1. For strength on centric compression: $\hat{Y}_1 = 9.841z_1 + 22.083z_2 + 7.597z_3 - 0.390z_1z_2 + 8.448z_1z_3 - 14.055 + 54.276z_1z_2z_3$
2. For bending strength $\hat{Y}_2 = 0.325z_1 + 0.5234z_2 + 0.2872z_3 + 0.1076z_1z_2 + 0.1763z_1z_3 - 0.1817z_2z_3 - 2.3221z_1z_2z_3$
3. For flow of water for mixing $\hat{Y}_3 = 0.204z_1 + 0.121z_2 + 0.350z_3 + 0.06518z_1z_2 + 0.169z_1z_3 - 0.1797z_2z_3 - 0.66851z_1z_2z_3$
4. For density $\hat{Y}_4 = 1549.48z_1 + 2032.4z_2 + 1417.2z_3 - 407.36z_1z_2 + 410.07z_1z_3 - 491.88z_2z_3 - 359.07z_1z_2z_3$

We remind the reader that for practical use of the obtained model, we need to decode the equations by substituting the dependencies between *natural* and pseudofactors in them. Now we are finally able to study in detail both numerically

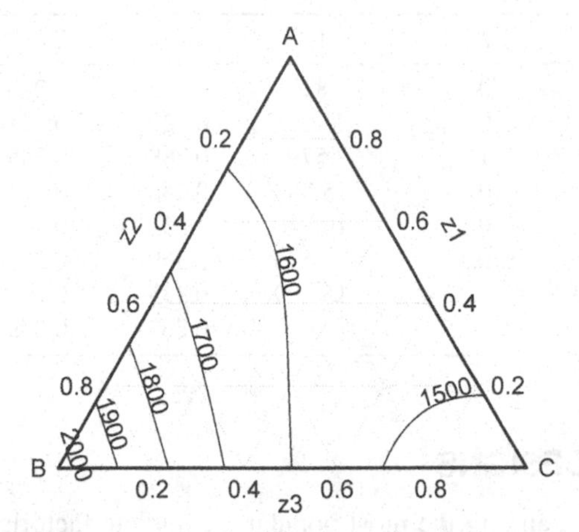

Figure 2.25 Density of Y_4 concrete samples, depending on the content of the components expressed by the pseudofactors- concentration of cement (z_1), sand (z_2), and addition of CBPB waste products (z_3).

Table 2.21
Model obtained for response Y_4.

Term	Estimate	Std.error	Statistic	P-value
z1	1549.2	15.499	99.954	0.0063689
z2	2032.4	15.501	131.11	0.0048554
z3	1417.2	15.491	91.486	0.0069584
z1:z2	-407.36	75.852	-5.3704	0.1172
z1:z3	410.07	75.93	5.4006	0.11656
z2:z3	-491.88	75.182	-6.5425	0.096558
z1:z2:z3	-395.07	474.38	-0.83281	0.55791

x1	x2	x3
0.333	0.333	0.333
0.250	0.750	0
0.143	0	0.857
0.291	0.541	0.167
0.238	0.166	0.595
0.196	0.375	0.428
0.242	0.361	0.397
0.219	0.329	0.452

(thanks to the previously presented models) and visually (thanks to the response surface reported) the behavior of the selected responses caused by the different concentrations of CBPB.

z1	z2	z3	Y1	Y2	Y3	Y4
1	0	0	9.81	0.325	0.205	1549.5
0	1	0	22.07	0.523	0.121	2032.6
0	0	1	7.579	0.289	0.349	1416.7
0.5	0.5	0	15.743	0.448	0.181	1690.1
0.5	0	0.5	10.77	0.35	0.32	1585.9
0	0.5	0.5	11.593	0.369	0.275	1599
0.333	0.333	0.333	15.509	0.335	0.228	1585.9
0.2	0.35	0.45	12.095	0.263	0.289	1589

2.10 OTHER DESIGNS

As reported by (10), among the most popular incomplete factorial designs, we can find the *Plackett – Burmann* designs. They are useful for gathering information on the main effects of the factors, which is the object of our study, but they do NOT provide information on their interactions.

The number of experiments required to perform is $4n$, where n can be $n = 1,2,3, etc.$ Due to its properties, a $4n$ Plackett (13) design can investigate $4n-1$ factors.

Due to the minimum number of experiments to perform in comparison with the number of factors to study, these design are very useful and so widely adopted when the number of factors of interest is less than the factors for a given design, since it allows the introduction of dummy factors. These factors, that have no chemical meaning, can help in determining the measurement error and to discriminate which factors are really contributing to our system and which not.

As an example, for $n = 1,2,3...$, the number of experiments to perform is $4,8,12$, so when we apply a design with 8, we can accommodate up to 3 dummy variables; when we use a design with 12, since we can study with it $4 \cdot 3 - 1 = 11$; if we need to study 9 variables (that we could not handle with 8 experiments), we can study up to 2 dummy variables. If we apply for the same problem a design with 16 experiments, we would have $(4n-1) - nvar = 15 - 9 = 6$ possible dummy variables to add, etc.

This is analogous to what we did before we can calculate ANOVA for the design and compare the t value for the dummy variables and the "real" one to discriminate their contribution to the system examined.

In the Appendix, we reported Plackett-Burman designs for n = 2, 3, 4, and 5.

As highlighted in the introduction of "A Brief History of Statistical Design" of the first chapter in (11), in the 60s, Kiefer proposed a formal approach to selecting a design based on specific objective *optimality criteria* (i.e., in the early stage, the optimality criteria was the best possible precision that model parameters estimated could give).

The subsequent development of computer tools lead to the advances in the field of *optimal design*.

As reported in (8), while dealing with multivariate ordinary least squares, the performance of estimates is judged by some properties of the *asymptotic variance-covariance matrix* $C^{-}1$ and more in details:

- D-optimal if it *minimizes the generalized variance* of the parameter estimates, that is, $det(C^{-1})$ is as small as possible
- A-optimal if it *minimizes the average variance* of the parameter estimates, that is, *trace* C^{-1} is as small as possible for the A-optimal designs (the *trace* of the matrix)
- E-optimal if it *minimizes the variance* of the least well-estimated parameter, that is, the *largest eigenvalue* of C^{-1} is as small as possible
- G-optimal if it *minimizes the maximum prediction variance* over all possible choice sets of a given size
- V-optimal if it *minimizes the average prediction variance* over all possible choice sets of a given size

While applying Design of Experiment, the reader will get acquainted with the name of *Genichi Taguchi*, who had a significant impact in this field. His approach was based on what he called robust parameter design using fractioned factorial design, orthogonal arrays, and novel statistical methods. *Taguchi methods* generated discussion and controversy, and details can be found again in (11). Indeed, their widespread adoptions have lead to positive outcomes.

As a reference on the topic, you can have a look at the work of Roy, "A primer on the Taguchi Method" (17).

2.11 FAQ

Q. How should I select a design for my experiment?

A. This is probably THE most frequently asked question.

A good answer is given by (15) (Chapter 5 page 115). The first thing to do when choosing a design is to focus on the numbers of factors that we need to study and what is the previous knowledge we have on the system we are going to characterize. If we have many factors to screen and little knowledge of the process in order to optimize it, we need to perform the identification of the most important variables. The following step involves the use of fractional factorial design, and eventually Plackett and Burman. After a first screening or in the case of a few factors to study, we can use a complete factorial design. At this point after optimizing the previously defined conditions, our objective will be to experimentally confirm the results obtained by the analysis of a previously computed response surface and its validation.

Q. How should I define the problem I am studying?

A. A very good workflow that will help you to gather correct answer is described by (7) (page 392) in the Selection of the best experimental strategy chapter. Defining the problem should include the description of the phenomenon, the definition of the responses in exam, the selection of factors to study, and finally defining the limits of the experimental domain.

Q. Is it necessary to know the experimental error?

A. Before answering this question I would remind the reader of the definition of *replicates*. Quite often I've seen that *performing a second read* of an instrumental measurement on a sample, using the same sample for a *second measurement* and preparing a *second sample* are considered to give the same information on the variance of a system. That, of course, is not true.

Only in the latter case, we can build a set of samples for testing the model built. Also, the following hypothesis should be checked:

- Normal distribution of the errors
- Lack of systematic errors
- Check homoscedasticy of the experiments error
- Independence of the variables
- Check from previous knowledge that the degree of the model is in accordance with the laws followed by our system (i.e., if we already know that in a chemical system, the response has a quadratic dependence with the temperature, it will not be correct to build a first degree model)
- Check if the nature of the factors in the system is additive before applying designs that require this condition (factorial design are additive and if this is not verified the effects and interactions will be false)

When we can perform repetitions of the measurements we should always get this chance in order to enhance the quality of the knowledge of the experimental error of the system.

Q. Why is it important to perform multiple measurements at the central point?

A. This is due to the fact that it can give us information about whether the agreement of the model with a first degree or higher (in our case mainly second degree) is valid.

2.11.1 EXERCISES

The reader can download the data-set based on the work from (12), shown in Table 2.11.1, and perform the following calculations:

EMAGMA	CE	Screw.speed	Block	Impact.strength
10	0.00	100	1	34
10	0.00	200	1	52
10	0.25	100	1	40
10	0.25	200	1	70
10	0.50	100	1	50
10	0.50	200	1	70
15	0.00	100	1	50
15	0.00	200	1	68
15	0.25	100	1	92
15	0.25	200	1	104
15	0.50	100	1	93
15	0.50	200	1	118
20	0.00	100	1	58
20	0.00	200	1	152
20	0.25	100	1	138
20	0.25	200	1	170
20	0.50	100	1	125
20	0.50	200	1	200

- Code the variables in order to apply a regression model
- Perform a multivariate regression using a first order model
- Perform a multivariate regression using a second order model

The reader will also find in the Appendix the tables for performing different kinds of designs of experiments, and also how to create scripts, using R software to build them.

2.11.2 REMARKS

At the end of this chapter, I would like to make a few remarks about the application of DoE.

The first question you would probably ask yourself after reading this chapter is: Why is DoE not widely adopted?

- A common misunderstanding is that if I apply DoE to my problem, I would consequently **replace** the previous **non-statistical** knowledge I acquired. This is absolutely **NOT** the case. Non-statistical knowledge is indispensable in selecting the factors, most suitable number of replicates, analyzing the variance, and interpreting the results obtained.
- Another common misunderstanding is that experimenter tends to *overestimate* the power of DoE.

A correct design of experiment is not a silver bullet for avoiding considering wrong factors, rediscovering the obvious, and for not making mistakes during the experimental part.

- Keep it simple.
  Simple screening designs are very powerful in selecting good candidate factors for a DoE performed to obtain more answers. Statistical designs offer a wide range of DoE design, but it does not mean that sophisticated statistical techniques give you better results than simple ones (Occam's razor).
- Finally, a good tip is to use an iterative approach to your problem. Start using a screening design, calculate a model, validate it, optimize your design in a subsequent step, validate it, and eventually restart, considering changing the factor selection performed.
  It will be difficult to know at the beginning of your experiment, to select the correct factors and levels, and also to miss some practical aspects of your experiment. Invest your resources wisely, considering that making only one run of your experiment rarely happens.

2.11.3 SUGGESTED ESSENTIAL LITERATURE

Two very useful complete reference books are "The theory and design of experiment" by Cox (4) and Montgomery's "Design and Analysis of experiments" (11).

A simple, focused approach on Design of experiment is given by Brereton in his introductory text "Applied Chemometrics for Scientist" (3).

As an additional source of case studies, the reader can also consider the book by Rodrigues "Experimental Design and Process" (14).

A comprehensive text on mixture design is Smith's "Experimental Design for Formulation" (18).

In order to get acquainted with R packages dedicated to the Design of experiments, the reader can rely on the book by Lawson, *Design and Analysis of Experiments with R* (9).

Bibliography

1. Valentin Amrhein, Sander Greenland, and Blake McShane. Scientists rise up against statistical significance. 567(7748):305–307, mar 2019.
2. Russell R. Barton. *Graphical Methods for the Design of Experiments.* Springer New York, 1999.
3. Richard G. Brereton. *Applied Chemometrics for Scientists.* John Wiley & Sons Inc.
4. D. R. Cox and N. Reid. *The theory of the design of experiments.* Chapman and Hall/CRC.
5. Valeriy Ezerskiy, Natalia Vladimirovna Kuznetsova, and Artem Denisovich Seleznev. Evaluation of the use of the CBPB production waste products for cement composites. *Construction and Building Materials*, 190:1117–1123, nov 2018.
6. Antonio Faria Neto and Antonio Fernando Branco Costa. Optimization of Pultrusion Process Parameters via Design of Experiments and Response Surface. *Quality and Reliability Engineering International*, 32(3):1265–1274, apr 2016.
7. J. Goupy. *Methods for experimental design: principles and applications for physicists and chemists.* Number v. 12 in Data handling in science and technology. Elsevier, Amsterdam ; New York, 1993.
8. Klaus Hinkelmann. *Design and Analysis of Experiments.* Wiley-Interscience.
9. John Lawson. *Design and Analysis of Experiments with R.* Chapman and Hall/CRC, 2014.
10. N.Miller Miller and Jane C. Miller. *Statistics and Chemometrics for Analytical Chemistry.* Prentice Hall, 2010.
11. Douglas C Montgomery. *Design and Analysis of Experiments Eighth Edition.* 2012.
12. Vidhya Nagarajan, Amar K. Mohanty, and Manjusri Misra. Reactive compatibilization of poly trimethylene terephthalate (PTT) and polylactic acid (PLA) using terpolymer: Factorial design optimization of mechanical properties. *Materials & Design*, 110:581–591, nov 2016.
13. R. L. Plackett and J. P. Burman. The design of optimum multifactorial experiments. *Biometrika*, 33(4):305–325, 06 1946.
14. M. Isabel Rodrigues and Iemma A. Francisco. *Experimental Design and Process Optimization.* CRC Press, 2014.
15. Maria Isabel Rodrigues and Antonio Francisco Iemma. *Experimental design and process optimization.* CRC Press, Taylor & Francis Group, Boca Raton; London, 2015. OCLC: ocn902837994.
16. Y. Rostamiyan, A. Fereidoon, A. Ghasemi Ghalebahman, A. Hamed Mashhadzadeh, and A. Salmankhani. Experimental study and optimization of damping properties of epoxy-based nanocomposite: Effect of using nanosilica and high-impact polystyrene by mixture design approach. *Materials & Design (1980-2015)*, 65:1236–1244, jan 2015.

17. Ranjit K. Roy. *A Primer on the Taguchi Method*. Society of Manufacturing Engineers, 1990.
18. Wendell F. (Wendell Franklyn) Smith. *Experimental design for formulation*. Society for Industrial and Applied Mathematics, 2005.

3 Pattern Recognition

Objectives:
Learn basic concepts of:

- correlation between variables
- principal component analysis
- the importance of preprocessing
- classification techniques

3.1 INTRODUCTION

As reported by Theodoridis (15), *pattern recognition*

> *is a discipline whose goal is to classify objects in categories or classes and its wide applications range from machine vision, character recognition, computer-aided diagnosis, speech recognition, etc.*

For each sample belonging to these application fields, we can describe a set of *features* that we can gather in a *feature vector* to be used to describe a *class*. After determining the features to examine and measure their values, we will be able to apply different *classification techniques* in order to discover the relations among our sample population.

Material science, due to the vast amount of characterization techniques employed and to its focus on studying correlations in set of samples, can greatly benefit to the application of pattern recognition.

Summarizing what we've presented so far, after *defining the goal* of our research and *identifying the variables* involved in the processes in exam, we *prepared an experimental plan by means of design of experiment*, and *performed the required analyses*, and *developed models* of our systems to *analyze the results obtained*

The last steps reported involves answering the following questions:

- Was our sample statistically significant?
- How did the variables influence our system?
- Can we *correlate* the answers obtained from our system to the selected variables?
- Can we *classify* the samples according to their properties?

Pattern recognition can help in all these tasks and finding its natural application in:

- Analysis performed with techniques that rely on a *huge number of factors*, i.e., all spectroscopy techniques
- Multivariate analysis, i.e., analysis where our final data set is obtained by merging together different analytical techniques used to create a *multivariate profile* of our samples
- Systems where we need to screen a huge number of products according to their final properties that depend on a wide range of factors, i.e., *libraries* of substances obtained via combinatorial techniques
- Screening of *existing* libraries of substances thanks to selected physico-chemical properties

In this chapter we will introduce a few essential techniques related to pattern recognition with the help of selected examples.

3.2 VARIABLE CORRELATION

As already reported in the introduction, we need to select the *factors* that we would like to use as *features* for creating our *feature vectors* in order to classify our samples.

We've already introduced in the previous chapter the concept of *correlation in errors* due to uncontrolled variations, but we need to remind the reader that *correlation* can also refer to *variables*.

A practical definition of correlation can be found in (16), where it is reported that: "researchers compute correlation coefficients when they want to know how two variables are related to each other".

As the reader probably already knows, there can be more than one definition of correlation. One of the most cited and used in practice is the *Pearson correlation coefficient*, also called the *product moment correlation coefficient*, written as ρ for a *population* and *r for a sample*. It can be calculated using the following formula:

$$r = \frac{SS_{xy}}{\sqrt{SS_x SSy}} \tag{3.1}$$

where SS_x is the sum of squares of x, obtained as $SS_x = \sum_{i=1}^{n} (x-\overline{x})^2$, SS_y is obtained in the same way of SS_x, and SS_{xy} is obtained using the formula

$$SS_{xy} = \sum_{i=1}^{n} (x-\overline{x})(y-\overline{y}) \tag{3.2}$$

The *range* of r is (-1,1), where *0 indicates no relationship between the variables (and or factors), while increasing (and decreasing values) represent a stronger (positive or negative) relationship between the variables and/or factors.*

Using the definition of *correlation* we just introduced, we will calculate the correlation of the **data-set PLA** reported in Chapter 1 based on (9), as reported in Figure 3.1.

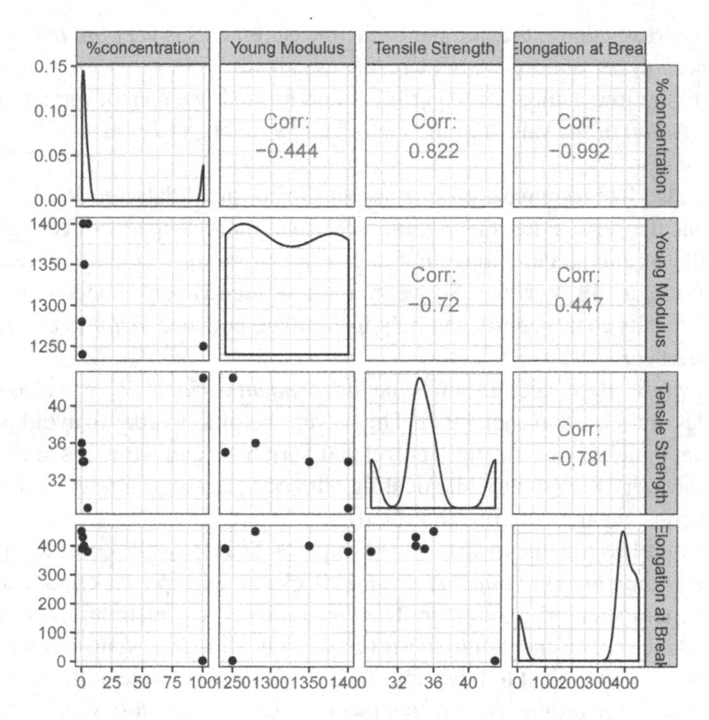

Figure 3.1 Correlation matrix of data-set PLA (9).

How can we interpret the results obtained?

The higher the correlation number, the higher the variables are correlated to each other, so starting from the first numerical value reported, we see that:

- *Content of PLA* is less correlated with the Young's modulus (−0.426) in comparison with the tensile strength (0.809) and the elongation at break (−0.993), for which we have *positive* and *negative correlations*, respectively
- *Young's modulus* is *negatively correlated* to the tensile strength (−0.695) while positively to the elongation at break
- We have a negative correlation of tensile strength and elongation at break (−0.766)
- A strong negatively correlation (almost 1) between the *percentage of the content of PLA* and the *elongation at break* also seems justified by the characteristic of the system, as also the negative correlations between tensile strength and young modulus and tensile strength and elongation at break.

The plot we presented also shows (in the left part and in the diagonal) the distribution of the correlation values for each sample in the form of a scatter plot (variable/factors vs variable/factors) and distribution plot (a curve that represents the distribution of the values of selected variables). What does this representation add to the previous analysis based only on the correlation values?

- as we already know, the number of the measurements is *very limited*, and so we need to be cautious with the results obtained
- due to the low number of experiments performed, we cannot see a *normal distribution* of the values of the factors and variable analyzed

I would also like to remark on one aspect of this small data-set. We know from the experimental details that there were three monitored properties (Young's Modulus, Tensile Strength, and Elongation at Break) on samples containing a different concentration % of PLA (wt%). So we monitored three variables depending on one factor. The results consequently can help us to relate how *one factor is correlated to one measured variable*.

In other words, the concentration is the *independent variable*. We can also see how the variables are related to each other. This is very helpful in order to avoid performing redundant analyses on strongly correlated variables, and we will see how these properties are very important while dealing with our data-set (think when instead of three properties, we measure one thousand).

At this point I would like to state a very important caveat while considering correlation: *Correlation does not imply causation*. A classic example reported in statistics is about the temperature in summer. There is indeed a correlation among the first variable and the second one, but it doesn't mean that if a nation wants to make its climate warmer, it is sufficient to buy a lot of ice cream! An example of the well-known fallacy *post hoc ergo propter hoc*, which means that if an event *followed* another one, it does not mean it was *caused* by it.

An example more close to material science is when we unwillingly *create a correlation* between two *independent data-sets*. As an example, we will make use of a simulated data-set made up of two **independent** series of measurements of the percentage concentrations (almost 10% with a standard deviation of 1%) of two elements in a metallic alloy. They are *uncorrelated* (as we imposed to be in our simulation), as it can also be seen in the scatter plot of Figure 3.2.

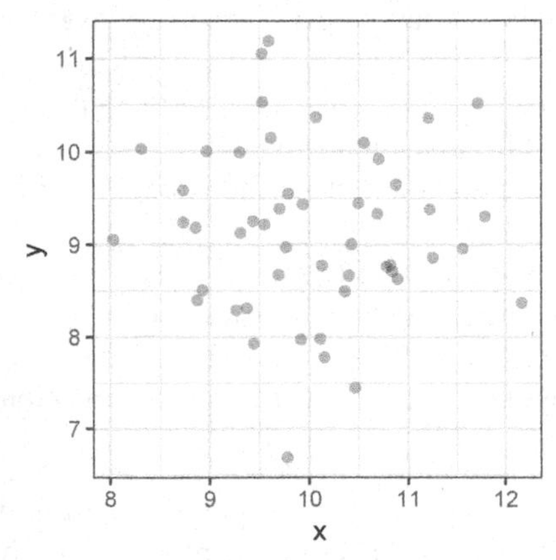

Figure 3.2 Scatter plot of the two independent series x and y representing two elements in a metallic alloy with a concentration around 10% and a standard deviation of 1%.

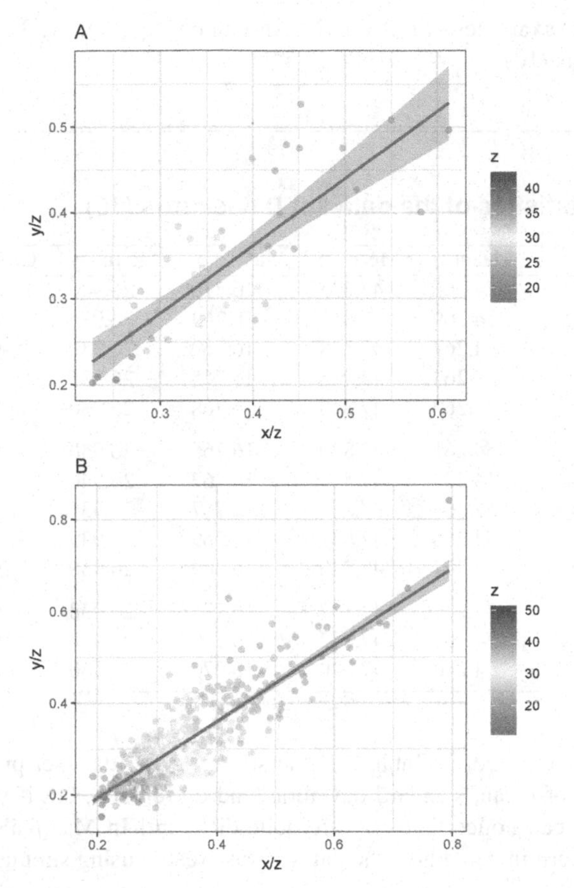

Figure 3.3 Correlation plot after dividing the two series of values of alloy elements for a third variable representing a bulk element with a mean value of 30% and standard deviation of 1% and 6%. Top simulation with 20 values, bottom 500 values.

What happens if we divide the two series for a third variable with a high standard deviation, as in Figure 3.3? In other words, for the content of another element in the alloy that has a higher concentration of mean value of 30% and standard deviation of 1% and 6%.

We can see that the *higher the values of the third variable and its associated standard deviation* we use, the *higher the correlation* between x and y we will find.

The explanation in this case is quite simple since after dividing for the third variable we are seeing the *correlation between the ratios* and not anymore the correlation between the original two variables!

3.2.1 DATASAURUS

Before introducing the PCA, we also need to remind the reader that two (or more) data-sets can present the *same results of descriptive statistics*, but look quite different.

Among the other examples of this are the Anscombe's quartet (1; 3), and the more recent Datasaurus (10).

Table 3.1
Descriprive statistics of the data-set Datasaurus (10).

Data-set	Mean(x)	Mean(y)	Std dev x	Std dev y	Corr(x,y)
away	54.266	47.835	16.770	26.940	-0.064
bullseye	54.269	47.831	16.769	26.936	-0.069
circle	54.267	47.838	16.760	26.930	-0.068
dino	54.263	47.832	16.765	26.935	-0.064
dots	54.260	47.840	16.768	26.930	-0.060
h lines	54.261	47.830	16.766	26.940	-0.062
high lines	54.269	47.835	16.767	26.940	-0.069
slant down	54.268	47.836	16.767	26.936	-0.069
slant up	54.266	47.831	16.769	26.939	-0.069
star	54.267	47.840	16.769	26.930	-0.063
v lines	54.270	47.837	16.770	26.938	-0.069
wide lines	54.267	47.832	16.770	26.938	-0.067
x shape	54.260	47.840	16.770	26.930	-0.066

As the reader can see, in Table 3.1, the subset of the data-set presents almost identical values of mean, standard deviation, and correlation, but if you plot them as in 3.2.1, you can notice that they are quite different! In Matejka's article (10), the reader can learn in detail how the dataset was created using simulated annealing methodology.

3.3 PRINCIPAL COMPONENT ANALYSIS

Principal component analysis is the workhorse of pattern recognition in chemometrics. Its aim is to reduce the *dimensionality* of a data-set while retaining as much *variance* as possible. It transforms the original variables in *uncorrelated linear latent variables* or *components* that can be extracted in order of decreasing content of *variance*.

As found in (17), "The direction in a variable space that best preserves the relative distances between the objects is a latent variable which has maximum variance of the scores. It is defined by a loading vector:"

$$p_1 = (p_1, p_2, ..., p_m).$$

The corresponding scores (projection coordinates of the objects, in chemometrics widely denoted by the letter t) are linear combination of the loadings and the variables . . . for objects *i*,...

$$t_{i1} = x_{i1}p_1 + x_{i2}p_2 + ... + x_{im}pm = x_i^T p_1$$

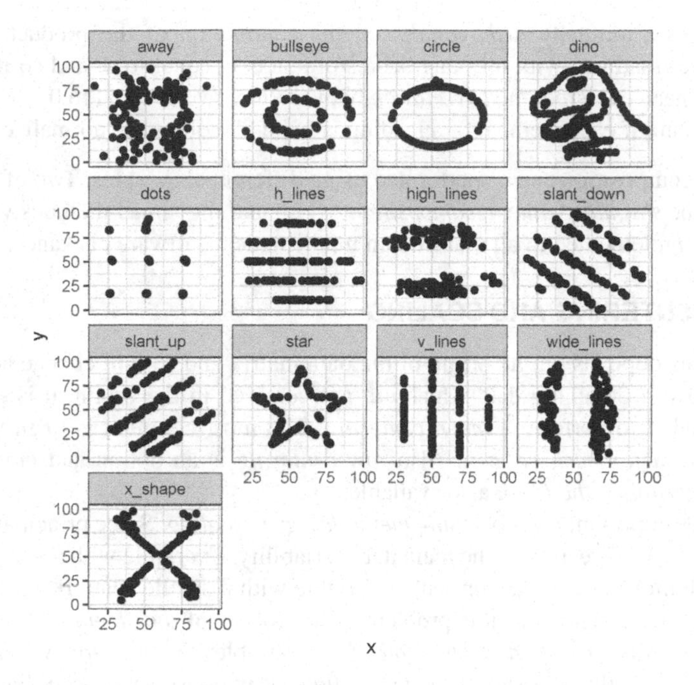

Figure 3.4 Scatter plot of Datasaurus data-set.

In matrix terms we can write

$$X = T.P + E$$

where

- X is the original data matrix
- T is the score matrix
- P the loadings matrix
- E is the error matrix

$$\underset{(I\times J)}{\mathbf{X}} = \underset{(I\times R)}{\mathbf{T}}\ \underset{(R\times J)}{\mathbf{P^T}} + \underset{(I\times J)}{\mathbf{E}}$$

If our matrix has dimensions $I \times J$ we can calculate a maximum number of J principal components. The *score* matrix T will consist of I column vectors while the *loading* matrix consists of J rows. Each column of the data matrix X is the vector x_j of observations on the *jth* variable.

It is useful to remind also a few properties of scores and loadings:

- they are mutually *orthogonal*, i.e., the summation of the products of a scores/loadings with all other scores/products is equal to 0, and so if they are mean centered, the *correlation* between any two is equal to 0
- the sum of each element of a loading is 1, i.e., loadings are normalized

The decomposition can be performed using different algorithms. Two of the most used are the *singular value decomposition (SVD)* and the *nipals* methods which are generally implemented in all mainstream mathematical software packages.

3.3.1 CENTERING AND SCALING

PCA results depends on the origin of the data matrix and so will change according to it. To avoid obtaining data which are not centered in the origin, it is generally recommended to perform a *mean centering*. The most common is to *subtract the column mean for each column*. Also *row centering* is an option and *more robust mean-centering methods* are also available.

Another important *preprocessing methodology* is *scaling*. Since principal components follow the direction of the main data variability, a variable with a a larger scale will weigh more in comparison with a variable with a smaller one. If we *autoscale* the variables we can avoid this problem. The process of *autoscaling* is performed using the *empirical standard deviation* of the variables or more *robust versions* of it. Keep in mind that if a variable contains *noise* it will also scale according to your preprocess.

Diagnostic plots for PCA

The *scree plot* is a plot that represents the cumulative sum of the variance of the data set according to the number of components extracted.

Score plots are scatter plots that report the scores obtained after performing PCA. They are useful for spotting outliers and clusters in the sample analyzed.

Loading plots are scatter plots reporting the loadings obtained from the PCA calculations, and they highlight the following properties

- if variables are *positively correlated* they will be *grouped together*
- if variables are in *opposed quadrants* they are *negatively correlated*
- the more *distant* the variables are from the origin (reported as arrows), the *better the quality* in representing the original variables

The scale of the score plot depends on *the original variables* while the loadings in the variable correlation are *standardized* (their maximum value is 1, represented by the circle with the origin in the center of the plot).

Another way of representing the results is to "merge" the *score plot* and *loading plot* in a plot called *biplot* (see 3.8), where both scores and loadings are reported. Its interpretation can be made in the say way as presented for the separate plot.

3.3.2 ALGORITHMS FOR PCA

We've seen that we can write scores as

$$t_{i1} = x_{i1}p_1 + x_{i2}p_2 + ... + x_{im}pm = x_i^T p_1$$

and also that we can define the first score as:

$$t_1 = x_{11}b_1 + ... + x_m b_{m1}$$

with unknown coefficients

$$b_1 = (b_{11}, ..., b_{m1})^T.$$

We are searching for the max variance of t_1 under the condition $b_1^T b_1$. The same goes for $t2$ considering both the condition $b_2^T b2 = 1$ and the orthogonality $b_1^T b_2 = 0$ and so on. We can then write

$$Var(t_j = Var(x_{11}b_1 + ... + x_m b_{m1}) = b_f^T Cov(x_1, ...mx_m)b_j = b_j^T \sum b_j$$

for $j = 1, ...m$ under the constraints $B^t B = I$.

This is a maximization problem under constraints that can be solved by writing it as Lagrangian expression and calculating its derivative with respect of b_j and setting the result equal to zero obtaining

$$\sum b_j = \lambda_j b_j$$

for $j = 1, ..., m$ which is known as EIGENVALUE PROBLEM, meaning that the solution for the unknown parameters is found taking for b_j the EIGENVECTORS of $\sum$ and for λ_j the corresponding EIGENVALUES. The problem can be solved using the algorithm SINGULAR VALUE DECOMPOSITION (SVD) or computing PCs as a sequential procedure where PCs are computed sequentially, maximizing the variance while using the orthogonality constraints. In brief the step of the *Nipals* algorithm are the following:

1)	$X(n \times m)$	X mean centered
2)	$u = x_j$	Initialize a score vector choosing j randomly
3)	$b = X^T \cdot u$ $b = b/\|\|b\|\|$	Calculate a first approximation of the vector and normalize it
4)	$u* = X \cdot b$	Calculate an improved score vector
5)	$u_\Delta = u * -u$ $\Delta u = u_\Delta^T u$	Check if there is an improvement or go to 6)
6)	$u = u*$	Substitute the previous score 2) with the improved one in 4)
7)	$t = u* \; p = b$	Store the score and loadings. Stop if we reached the desired n of components
8)	$X_{res} = X - u \cdot b^T$	Calculate residual matrix
9)	$X = X_{res}$	Substitute X with X_{res} and continue the calculation of PCA

As a final note the name Nipals is the acronym for "Nonlinear Iterative vartial Least Squares". It was developed by H. Wold at first for PCA and later-on for PLS. Nipals is also a mountain in Sweden, which is the motherland of Wold.

3.3.3 DATA-SET ELE: EXAMPLE OF PCA APPLIED TO A DATA-SET OBTAINED VIA ELECTROPHORESIS CHARACTERIZATION

We will present an application of PCA using a data-set derived from the work of Yucel et al. (19).

In the original work the authors used a capillary electrophoresis system to characterize the quarry marbles from three regions of Turkey. PCA was employed (successfully) to see if the elemental composition could be used to group the marble and eventually which elements were the most informative to distinguish among them.

The data was obtained by measuring the concentration of *10 metal ions* in *25 marble samples from three regions of Turkey, characterized using capillary electrophoresis*, as reported in Tables 3.2 and 3.3.

The first step was to standardize all the data since we want all the elements to have an equal influence over the results.

With the help of a R package (as reported in the appendix), we then performed a PCA, saved the model, and reported the most common diagnostic plots.

The *scree plot* (Figure 3.5) shows that two components represent almost 68% variance of the problem.

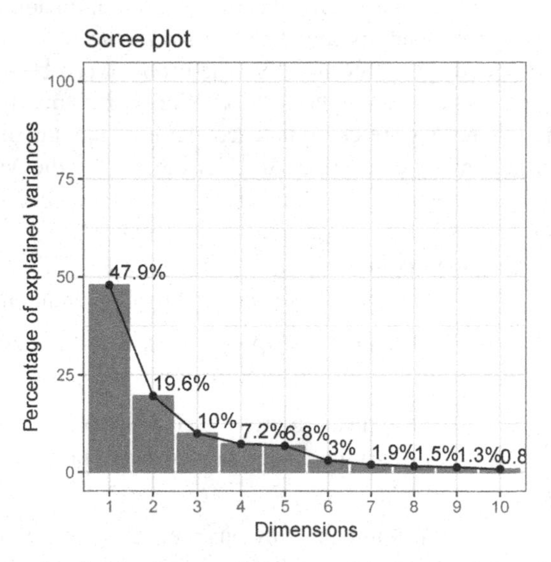

Figure 3.5 Scree plot for the data-set (19) reported in Figures 3.2 and 3.3. The first two components represent almost 68% of the variance of the dataset.

We report now the *PC1 vs PC2* plot. In Figure 3.6, samples marked with *T* are from *Thrace*, *A* from *Aegean*, and *M* from Marmara. In the figure the bigger symbols

Table 3.2
Data-set derived from the work of (19), depicting concentration of *10 metal ions* in *25 marble samples* from three regions of Turkey, characterized using capillary electrophoresis.

id	Ag	Fe	Cr	Mn	Cd
S1	2.41± 0.02	3.7± 0.8	0.60± 0.02	1.1± 0.3	0.08± 0.03
S2	2.2± 0.2	5± 1	0.43± 0.02	1.12± 0.02	0.07± 0.02
S3	6.02± 0.04	4.0± 0.1	1.70± 0.03	1.56± 0.03	0.84± 0.03
S4	1.36± 0.05	3.850± 0.008	0.70± 0.03	0.94± 0.02	0.50± 0.07
S5	6.75± 0.05	4.11± 0.02	1.00± 0.01	0.880± 0.002	0.44± 0.01
S6	0.67± 0.04	2.220± 0.005	0.75± 0.01	0.43± 0.01	0.12± 0.02
S7	5.23± 0.03	2.08± 0.05	0.26± 0.01	0.66± 0.04	0.265± 0.005
S8	9.6± 0.1	1.58± 0.06	0.440± 0.002	0.6± 0.1	0.47± 0.02
S9	10.9± 0.3	6.45± 0.04	0.68± 0.02	0.34± 0.04	0.370± 0.009
S10	10.1± 0.2	2.25± 0.06	0.48± 0.08	0.35± 0.02	1.38± 0.07
S11	8.98± 0.05	3.8± 0.4	0.7± 0.2	0.49± 0.02	0.42± 0.01
S12	3.80± 0.08	1.8± 0.2	0.34± 0.01	0.35± 0.04	0.42± 0.06
S13	2.76± 0.03	2.3± 0.2	0.42± 0.03	0.38± 0.04	0.42± 0.02
S14	10.0± 0.2	8.4± 0.3	1.69± 0.08	1.14± 0.05	0.96± 0.03
S15	12.3± 0.4	7.2± 0.4	0.80± 0.03	1.08± 0.07	0.78± 0.02
S16	6.2± 0.3	3.6± 0.2	1.06± 0.01	0.6± 0.1	0.7± 0.1
S17	12.2± 0.3	6.0± 0.1	0.94± 0.02	1.22± 0.04	0.94± 0.07
S18	10.55± 0.07	3.0± 0.1	0.46± 0.08	0.4± 0.1	0.37± 0.01
S19	11.7± 0.6	6.1± 0.6	1.00± 0.04	1.34± 0.08	1.0± 0.2
S20	9.113± 0.006	3.22± 0.05	0.555± 0.003	0.58± 0.01	0.715± 0.009
S21	4.9± 0.3	1.9± 0.2	0.52± 0.03	0.53± 0.05	0.73± 0.02
S22	10.2± 0.1	3.1± 0.2	0.48± 0.01	0.62± 0.02	0.700± 0.001
S23	1.5± 0.1	3.3± 0.3	0.54± 0.09	0.71± 0.08	0.40± 0.07
S24	10.3± 0.4	5.2± 0.5	0.70± 0.03	0.8± 0.1	0.8± 0.1
S25	1.2± 0.1	1.73± 0.01	0.30± 0.01	0.30± 0.02	0.26± 0.01

represent the average of the principal component of all the samples belonging to same category. The colors help in suggesting the similarities among the samples of the same category.

The plot of Figure 3.7 shows the weight of the original variables in the space of the calculated components.

In order to also have a visual interpretation to answer the question about how the selected variables correlate, we also report the correlation plot of the variables, as in Figure 3.9. The darker the color the higher (positive or negative) the correlation. The x represents variables which are not considered correlated using a threshold of 0.4. As we previously reported this is an hint that the variables are *correlated*. Again, a reminder that caution is necessary before taking the decision to avoid measuring only elements with no correlation among them. That is, due to the dimension of the

Table 3.3
Data-set derived from the work of (19), depicting concentration of *10 metal ions* in *25 marble samples* from three regions of Turkey, characterized using capillary electrophoresis.

id	Co	Pb	Ni	Zn	Cu
S1	0.36± 0.06	1.38± 0.02	0.832± 0.008	0.9± 0.1	0.9± 0.3
S2	0.27± 0.04	2.2± 0.3	0.66± 0.03	0.5± 0.1	0.86± 0.02
S3	0.60± 0.04	0.59± 0.09	0.66± 0.08	0.25± 0.02	1.455± 0.008
S4	0.38± 0.03	3.1± 0.4	0.66± 0.06	0.47± 0.06	0.67± 0.01
S5	0.7± 0.1	3.28± 0.06	0.72± 0.02	0.656± 0.005	0.85± 0.02
S6	0.24± 0.06	0.378± 0.009	0.204± 0.009	0.277± 0.008	0.545± 0.005
S7	0.206± 0.008	0.51± 0.02	0.387± 0.005	0.21± 0.01	0.526± 0.007
S8	0.18± 0.02	3.62± 0.03	0.765± 0.005	0.17± 0.02	1± 7
S9	0.25± 0.01	1.53± 0.02	0.60± 0.02	0.265± 0.005	0.583± 0.008
S10	0.43± 0.02	11.3± 0.5	0.47± 0.07	0.391± 0.002	1.1± 0.1
S11	0.278± 0.002	2.1± 0.2	0.36± 0.02	0.20± 0.02	0.49± 0.04
S12	0.40± 0.02	3.24± 0.03	0.28± 0.02	0.24± 0.03	0.63± 0.04
S13	0.345± 0.009	3.054± 0.008	0.25± 0.02	0.282± 0.002	0.55± 0.02
S14	0.605± 0.008	3.7± 0.2	0.50± 0.02	0.43± 0.02	1.13± 0.02
S15	0.40± 0.05	7.8± 0.7	0.32± 0.04	0.30± 0.04	0.52± 0.08
S16	0.57± 0.09	9.4± 0.2	0.31± 0.03	0.60± 0.01	0.63± 0.02
S17	0.62± 0.01	11.85± 0.07	0.38± 0.04	0.45± 0.02	1.067± 0.008
S18	0.138± 0.005	0.90± 0.03	0.158± 0.004	0.135± 0.003	0.18± 0.03
S19	0.7± 0.1	10.3± 0.5	0.49± 0.04	0.86± 0.04	1.2± 0.1
S20	0.30± 0.02	5.2± 0.1	0.40± 0.02	0.371± 0.009	0.67± 0.02
S21	0.23± 0.08	3.6± 0.3	0.26± 0.04	0.28± 0.04	0.37± 0.04
S22	0.5± 0.1	5.8± 0.2	0.298± 0.002	0.33± 0.02	0.69± 0.02
S23	0.36± 0.05	5.6± 0.1	0.357± 0.004	0.26± 0.04	0.537± 0.008
S24	0.72± 0.01	5.7± 0.2	0.38± 0.02	0.370± 0.002	0.57± 0.02
S25	0.25± 0.05	0.640± 0.008	0.150± 0.008	0.230± 0.001	0.40± 0.02

sample used, the absence of a training test, and the chemical aspects involved, I won't leave out any of the elements analyzed.

From the score plot we can see that

- *score plot* presents clustering for each of the groups
- looking at the average values of the first two pcs, we can discriminate among the three groups- T for *Thrace*, *A* for *Aegean*, and *M* for Marmara
- having a look at all the individual samples, we see that the specimens belonging to Thrace and Aegean can be discriminated from the ones from Marmara. The first two groups lay almost entirely in the I and II quadrant, while the Marmara ones are in the III quadrant
- the data-set does not include any suspicious outlier

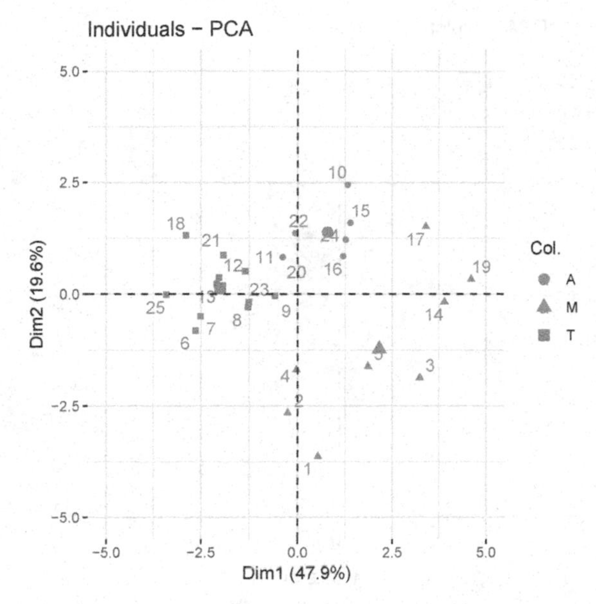

Figure 3.6 Plot for PC1 vs PC2 of data-set (19) reported in Figures 3.2 and 3.3. The first two components represent almost 67% of the variance of the data-set. Data was auto-scaled before performing PCA.

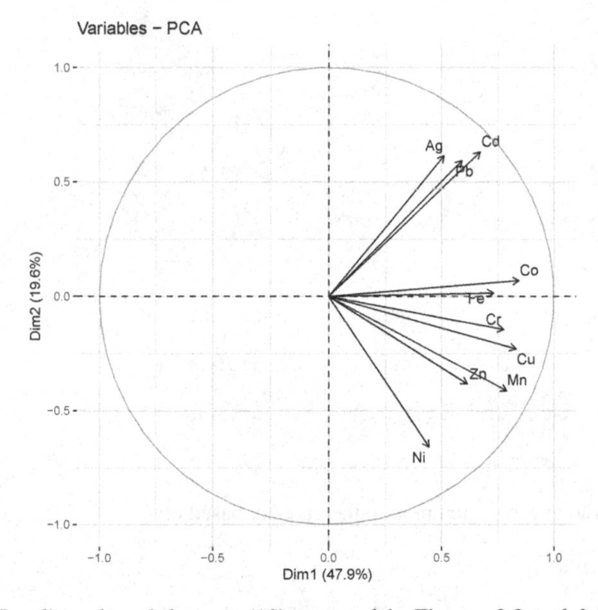

Figure 3.7 Loading plot of data-set (19) reported in Figures 3.2 and 3.3. The first two components represent almost 67% of the variance of the data-set.

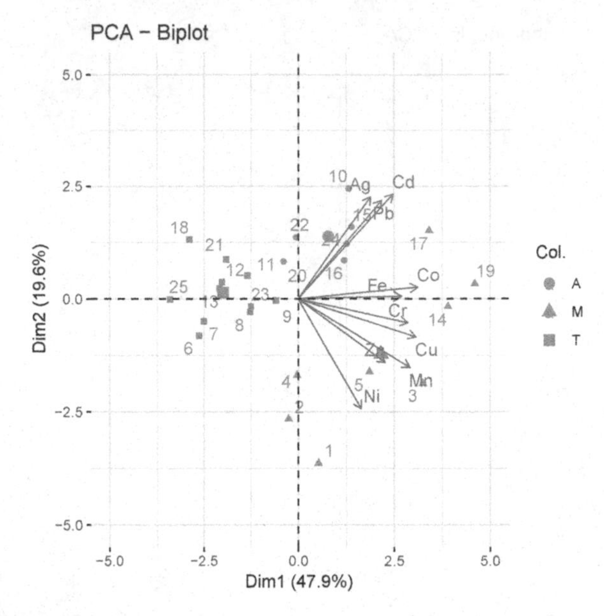

Figure 3.8 Biplot of data-set (19) reported in Figures 3.2 and 3.3. Both scores and loadings are reported.

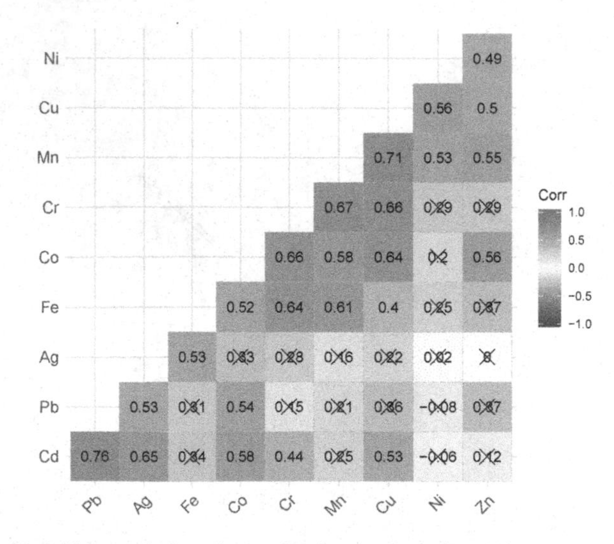

Figure 3.9 Variable correlation plot on the data-set based on (19).

From the loading plot we can gather the following info:

- all metal ions have similar weights in the calculation performed included between 0.6 and 0.8

- ion metals Ag, Pb, and Cd have similar weights while rotating our data-set. The same is also true for the following pairs- (Co, Fe), (Cr, Cu), and (Mn, Zn)
- ion metals Ag, Pb, and Cd are the variables that weigh more in order to discriminate the sample from Aegean from the other groups, while Mn, Zn, Cr, Fe, Co, Cr help us in discriminating the sample belonging to Marmara in comparison with Thrace and Aegean

The correlation plot highlights that the higher correlation occurs among (Cd, Pb), (Cu, Mn), and (Cr, Mn).

Summarizing:

- we can see differences among the specimens depending on their provenience
- we can highlight that the specimens belonging to the Thrace and Aegean can be discriminated from the ones from Marmara- the first two groups lay almost entirely in the I and II quadrant, while the Marmara ones lay in the III quadrant
- the data-set does not include any suspicious outlier
- the different content of ion metal can discriminate among samples

Now as an exercise we will have a look at the results of a principal component analysis on the same data-set *without scaling* in order to see what we would have obtained without enhancing the difference due to elements that are present in lower concentration in comparison to other elements.

In Figures 3.10 and 3.11, we report the screen and biplot, respectively.
The first difference that we notice in comparison with the previous analysis is in the value of explained variance. Now one component explains almost 71% of the variance as visualized in the *scree plot*.

Scaling also increased the power of differentiating among the groups. *Without scaling*, the score plot presents *no clear clustering*.

In the *loading plot* (and the biplot), we see that the concentration of Fe becomes the most important loading in rotating the matrix. The *biplot highlights* a lower contribution of the element Cd in separating the sample from Aegean from the other ones.

Due to the additional results gathered *without scaling*, in an analogous case to this one (a component with much higher concentration in comparison with the other one), I would proceed to create models with and without preprocessing to better describe the characteristic of the data-set.

3.3.4 DATA-SET ASPHALT: AN APPLICATION OF PCA TO ATR-FTIR SPECTROSCOPY

PCA technique is extensively used while analyzing spectroscopic data due to the high correlated nature of the variable in exam. We present now an application based on the work of (13), where the authors studied the developing of a method for rapidly

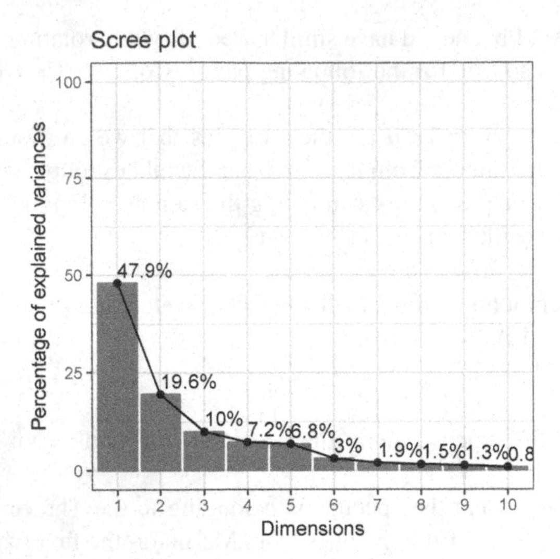

Figure 3.10 Scree plot for data-set (19) reported in Figures 3.2 and 3.3. The first two components represent almost 91% of the variance of the data-set.

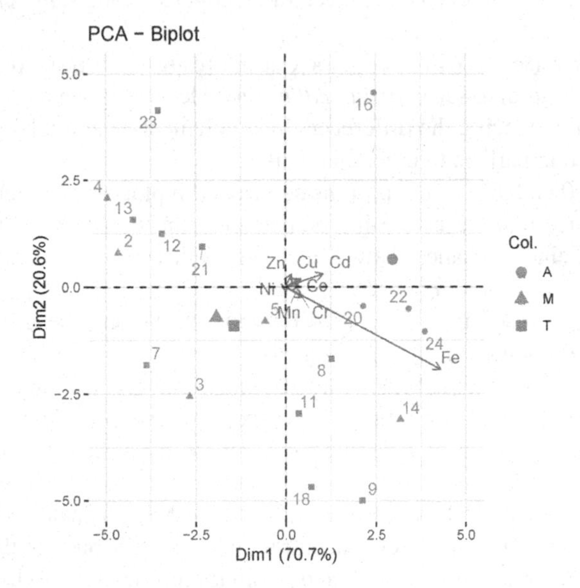

Figure 3.11 Biplot for data-set (19) reported in Figures 3.2 and 3.3. The first two components represent almost 91% of the variance of the data-set.

identifying oil sources and the quality of asphalt. The sources of the samples and their IDs are presented in 3.3.4.

The experimental matrix is composed of *26 spectra* recorded at wavelengths in a range of *600-300* $cm^{-}1$.

id	Source	Category
A1	Middle East	2
A2	South America	3
A3	Middle East	2
A4	South America	3
A5	Middle East	2
A6	Middle East	3
A7	Middle East	2
A8	Middle East	2
A9	South America	2
A10	Middle East	2
A11	South America	3
A12	Middle East	2
A13	South America	3
A14	Middle East	2
A15	Middle East	3
A16	South America	2
A17	South America	2
A18	Middle East	3
A19	Northwest China	3
A20	Northwest China	2
A21	Middle East	1
A22	Middle East	1
A23	south America	2
A24	Northwest China	2
A25	Middlc East	3
A26	South America	1

First of all, we apply a PCA after standardizing the spectra, considering the whole range of frequencies acquired. In this case, we apply an autoscaling, since we would like to "give the same weight" to all variables that correspond to the different chemical groups present in our sample. In Figure 3.13, we reported the scree plot, while in Figures 3.14 and 3.15, the score and loading plot are presented.

This time, in order not to present an overcrowded biplot, we separately reported the score plot and loading plot. The scree plot shows that the first two components explain almost 87% of the variance of the data-set.

The score plot lets us mainly define one cluster for category 2 (Middle east samples, I and IV quadrant) and 3 (South America, II and II quadrant), while the specimens belonging to the other category (Northwest China), due also to the small number of tests performed, do not individuate their own clusters (limiting the analysis to the first two PCs).

We've highlighted a selection of the variables in the loading plot. The frequencies highlighted refer to the characteristic adsorption group peaks of the substances in the exam, as reported by (13).

Considering our starting point, the result is quite good. Giving a first glance at the plot reported in Figure 3.12 it would be almost impossible to spot two different groups of analysis, but can we improve the result obtained?

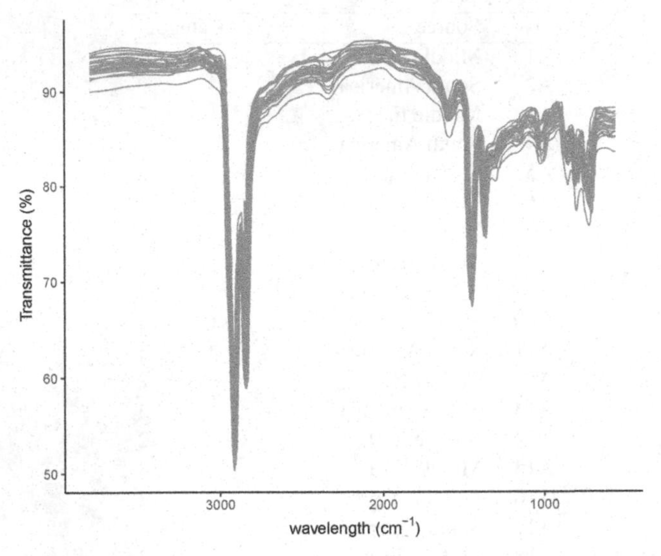

Figure 3.12 ATR-FTIR spectra for the samples presented by (13).

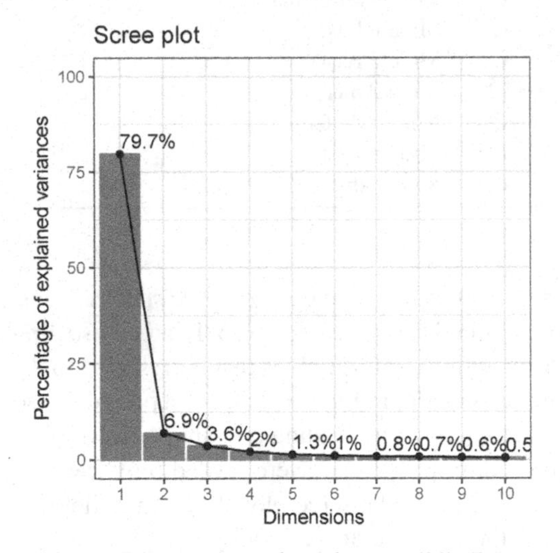

Figure 3.13 Scree plot of PCA performed on data-set (13). Data are autoscaled and no selection was performed on the variables.

This time, instead of considering the *whole spectra*, we will perform the PCA again, considering the *absorbance of the samples* in the exam at specific frequencies reported in Figure 3.4. The diagnostic plots for this model are reported in Figures 3.16 and 3.17.

We can notice that the variance explained increased (more than 90% explained by the first two variables), but the most important observation is that we can now separate all three categories of asphalts thanks to only the first two components.

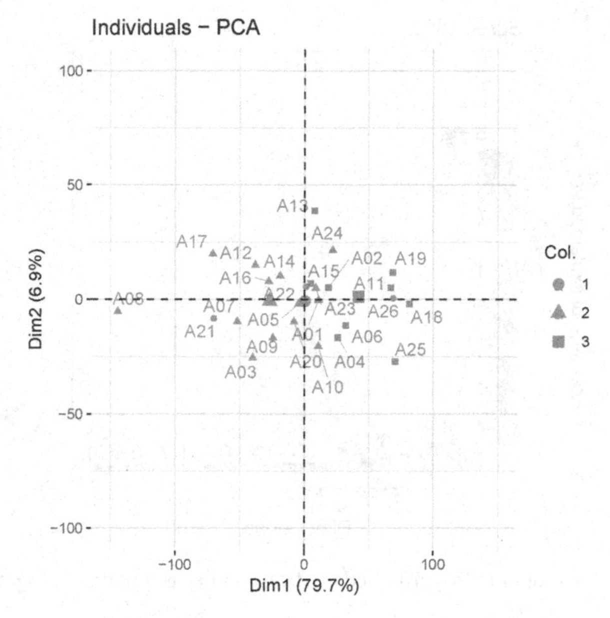

Figure 3.14 Score plot of PCA performed on data-set (13). Data is autoscaled and no selection was performed on the variables.

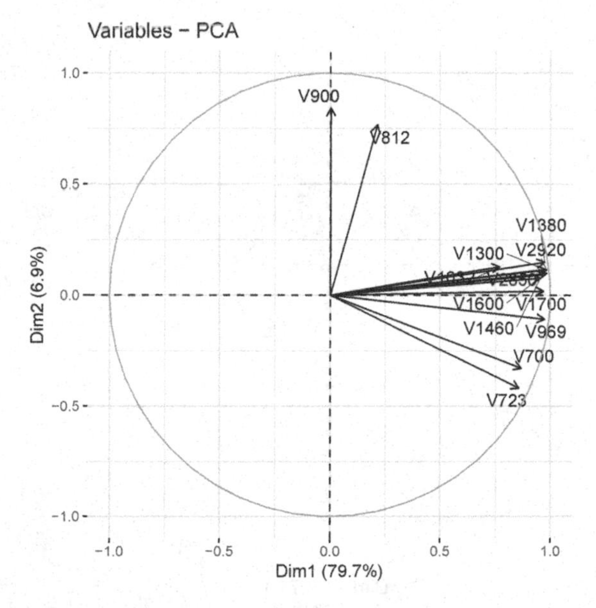

Figure 3.15 Loading plot of PCA performed on data-set (13). Data is autoscaled, and no selection was performed on the variables.

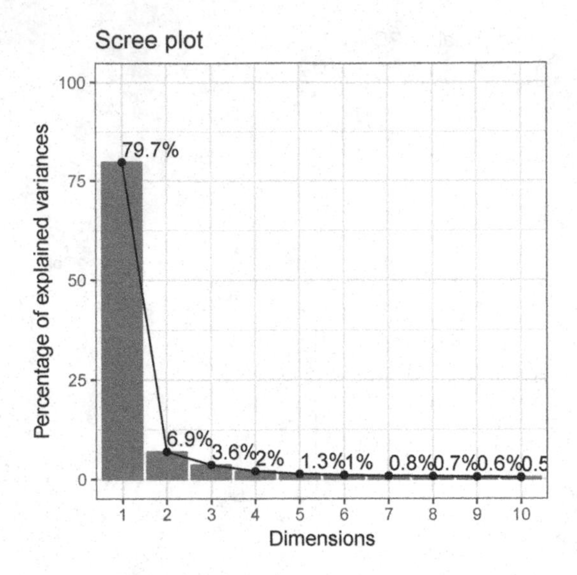

Figure 3.16 Scree plot of PCA performed on data-set (13) as reported in Figure 3.4.

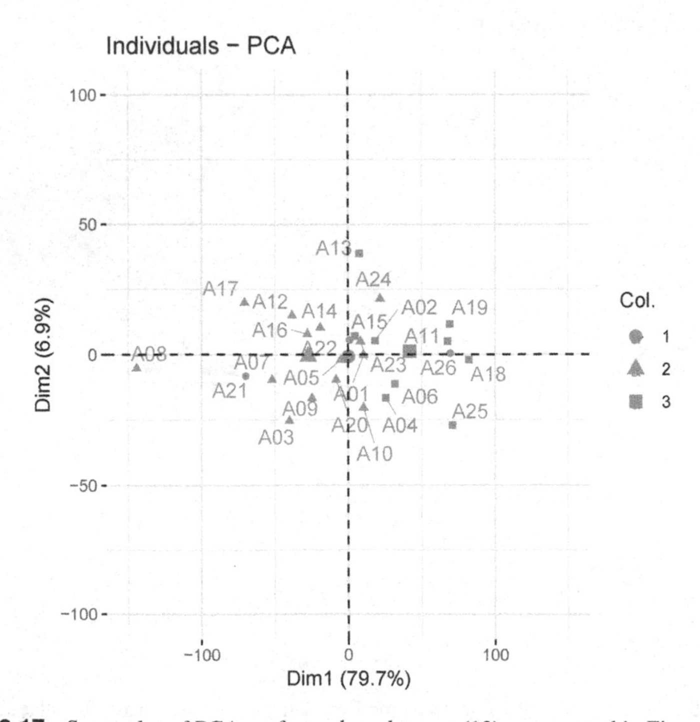

Figure 3.17 Score plot of PCA performed on data-set (13), as reported in Figure 3.4.

Table 3.4
Values of absorbances at selected frequencies presented by (13).

id	721	810	873	968	1030	1166	1375	1455	1599	1700	2850	2920
A1	79.14	81.85	84.26	87.11	85.14	85.88	75.78	68.80	87.58	91.36	59.51	51.08
A2	80.74	81.67	84.39	86.09	83.93	84.93	74.91	68.45	87.35	91.27	63.60	55.07
A3	77.71	80.27	83.44	86.18	83.98	84.94	75.35	68.72	86.97	90.81	59.96	52.46
A4	80.84	82.08	84.74	86.80	86.10	85.54	74.99	68.41	87.66	91.73	61.67	53.29
A5	78.36	81.23	84.12	86.85	84.61	85.66	76.50	69.80	87.72	91.14	60.71	54.28
A6	81.38	82.15	84.59	87.22	86.30	85.68	74.77	68.27	87.49	91.70	62.60	53.97
A7	77.69	80.09	83.56	85.74	83.03	84.42	75.13	68.76	87.20	90.35	60.96	53.10
A8	76.17	77.23	81.00	84.97	82.29	83.43	74.08	67.64	85.10	89.28	60.82	55.46
A9	78.43	81.27	83.68	86.57	84.49	85.29	74.92	67.86	87.04	90.93	58.47	49.83
A10	78.69	81.56	84.51	87.73	86.35	86.33	76.57	69.57	88.18	91.96	59.51	51.21
A11	82.56	83.39	85.41	87.21	84.98	85.91	77.26	71.68	88.18	91.26	67.38	61.02
A12	77.73	80.35	83.36	85.95	83.33	84.66	74.98	68.22	86.79	90.68	59.45	51.41
A13	78.25	80.40	83.79	85.98	82.57	84.65	76.43	70.32	87.72	91.31	64.78	59.69
A14	77.94	81.12	84.15	86.89	85.01	85.65	75.54	68.55	87.83	91.46	58.45	50.18
A15	80.03	81.70	84.41	86.67	84.84	85.28	75.31	68.85	87.50	91.15	62.17	54.10
A16	77.61	80.69	83.60	86.72	84.37	85.41	75.88	68.58	87.37	91.10	58.43	50.44
A17	77.27	79.82	83.33	85.75	83.68	84.47	74.48	67.84	86.91	90.44	59.72	51.84
A18	81.22	83.88	85.16	88.10	86.80	86.52	76.90	70.17	87.35	91.73	60.74	53.96
A19	81.87	82.79	84.49	87.52	86.56	86.00	75.80	69.46	87.20	90.83	61.90	54.30
A20	78.18	81.33	84.07	86.96	84.78	85.88	76.07	69.14	87.75	91.56	58.74	51.30
A21	84.78	88.21	88.53	90.14	89.81	89.08	78.50	71.67	90.26	92.29	61.69	53.23
A22	84.90	88.24	88.50	90.15	89.81	89.10	78.96	72.24	90.08	92.41	62.36	53.29
A23	78.93	81.82	84.64	87.36	85.20	86.33	76.95	70.10	88.25	91.85	59.27	52.17
A24	78.14	80.59	83.93	87.02	85.40	85.57	75.48	68.64	87.58	91.88	59.69	51.62
A25	82.70	83.75	86.09	87.75	86.61	86.84	77.17	71.05	88.74	92.29	65.24	56.59
A26	87.15	89.43	89.86	90.69	90.68	90.25	78.07	72.18	91.63	94.06	64.88	56.64

So now the obvious question is: "Is it always the case to select only the variables of the most common adsorption peaks before performing a PCA"?

The answer is not so immediate. In this particular case, using a selection of variables was sufficient to rule out the noise present in the data-set and increase the discriminating power of PCA, but is it always like this? We need to pay attention when making assumptions that leave part of our data-set. First of all we need to think about the preprocessing of our spectra. When we auto-scale (a very common procedure already presented), every variable is going to have the same "weight" in the analysis as the other, but contemporaneously we also scale up and increase the noise present in the spectra.

Also, the selection MUST occur only for frequencies (or interval of frequencies) that FOR SURE contain information about the substances we are investigating. If we select a range of frequencies where the techniques are not able to characterize any functional group, the results will be that we would have less variables to work on, and thus there would be a reduction in the information available for calculating a model, decreasing the discriminatory power of the PCA.

We also, of course, need to use all the spectroscopic knowledge we have on the problem helping us in making meaningful choices.

In our case, we knew which ones were the frequencies to select, since they are characteristic of the investigated groups present in the specimen. Thus, it was a legitimate and a sensible choice to leave other frequencies out.

Also, in order to know if the models obtained are trustworthy, we should test them. How can we do that? The best thing is to *compare the projected values* of unknown samples and compare them on another sample where the specimens were characterized in a *totally independent run*. The latest is defined as *test set*. The comparison is generally the *sum of the squared residuals*. It can be called also *residual sum of square* and in this case (where we can use an eternal set) as *sum of squared errors in prediction*. If we cannot make use of two independent runs, we can still calculate the quality of our model by splitting the set to a *training set* and a *test set*, which contain, respectively, about 70% and 30% of the samples. If the data-set does not contain a sufficient number of specimens to create a training and a test set (that for practical reasons, we can approximate in 15), or in addition to splitting it, we can also recur to *cross validation*, where a number of samples (1 to 5) are iteratively left out, and at each iteration, the model obtained is tested for its predictive capability.

3.3.5 DATA-SET PCAMIX: PCA APPLIED TO BINARY CHEMICAL MIXTURES AT TRACE LEVELS

We've seeen in the previous examples how PCA can be used in the spectroscopic field and we will present an additional example to answer the following questions. Can PCA be used to analyze elements in trace quantities? Can we also use it in cases where we need to characterize mixtures?

Our data-set is based on the work of (5), where the authors demonstrated the validity of the triangle-rule and balance-rule in estimating the composition of ternary and binary mixtures of methyl orange (MO), methylene blue (MB), and crystal violet (CV) based on PCA performed on surface-enhanced Raman Scattering spectra.

The sample composition is shown in Table 3.5.

Table 3.5
Sample composition of all concentrations expressed as percentages-methyl orange (MO), methylene blue (MB), and crystal violet (CV).

id	MO %	MB %	CV %
A1	54± 4	46± 4	...
A2	44± 6	...	56± 6
A3	...	45 ± 5	55± 5
A4	74 ± 1	26 ± 1	...
A5	...	76± 1	24± 1
A6	38 ± 4	38± 5	24± 10
A7	66± 1	16± 1	18± 1
A8	22± 5	65± 2	13± 4

First of all, we report the Raman spectra of the pure substances, and then mixture samples in Figures 3.18 and 3.19.

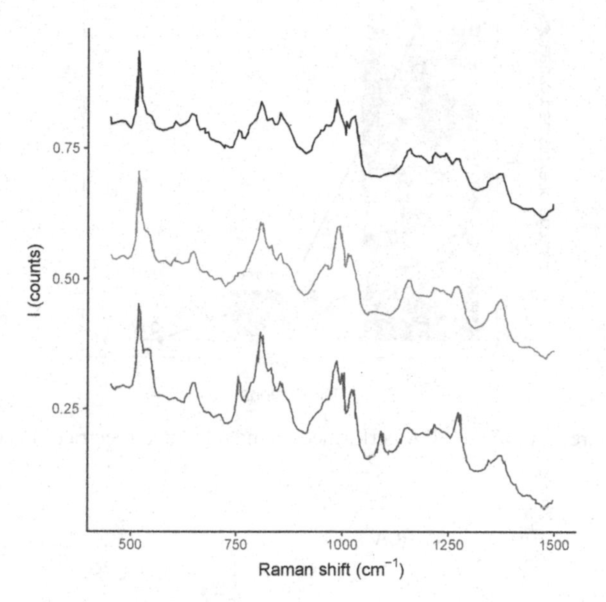

Figure 3.18 Spectra of sample references based on (5), with starting lower values of I for Crystal Violet, Methyl Blue, and Methyl orange.

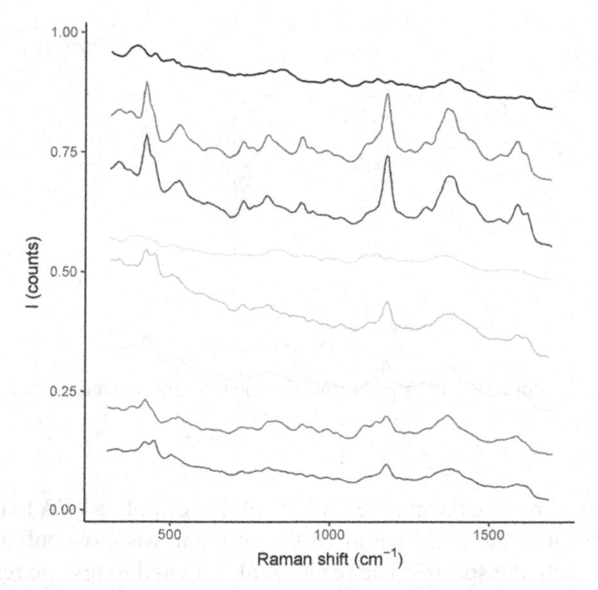

Figure 3.19 Spectra of mixtures based on (5), starting from higher values A1-A8. Compositions are reported in Table.

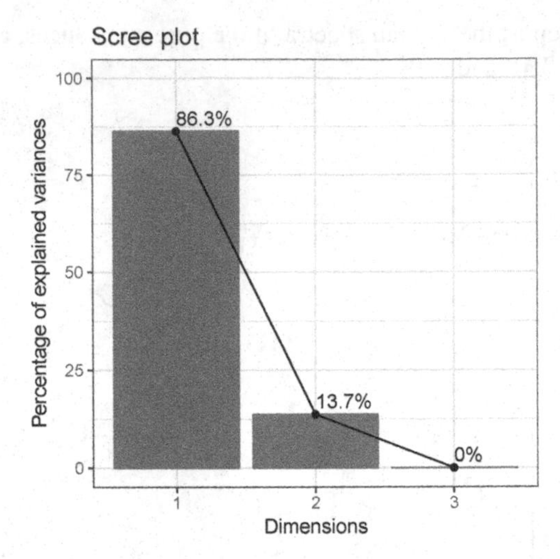

Figure 3.20 Scree plot of the PCA performed on the reference spectra of the data-set based on (5).

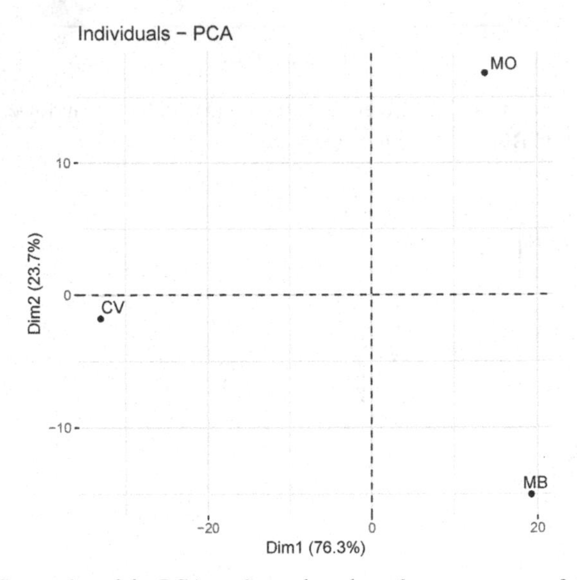

Figure 3.21 Score plot of the PCA performed on the reference spectra of the data-set based on (5).

Since we want to reduce the number of variables, we apply a PCA to the spectra of the pure components. Due to the nature of the problem, we show only the workflow centering and scaling the spectra. The reader is also invited to test the results without applying any preprocessing to the spectra. We can find that two components are more than enough to represent the variance of the problem, as reported in Figures 3.20 and 3.21.

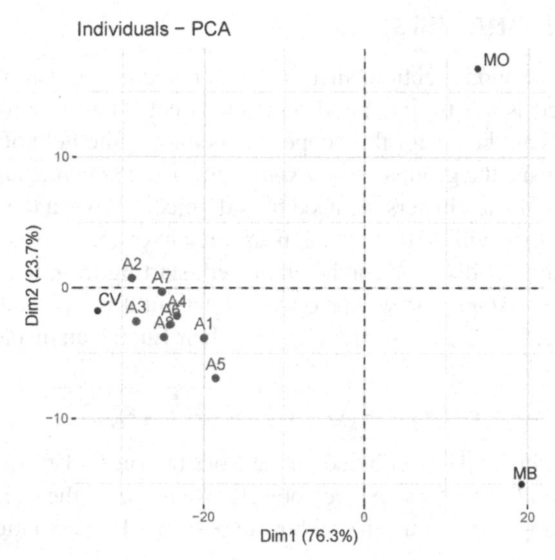

Figure 3.22 Score plot of the PCA performed on the reference spectra of the data-set based on (5).

At this point, in order to perform the calculations necessary to find the composition of the mixture applying the triangle rule for a mixture, instead of reapplying a PCA *we can project the spectra of the mixture samples in the space we've just determined for the reference samples.* The results are highlighted in figure.

The same kind of projection is used while we need to insert new data in an already calculated model. This kind of process finds many applications as an examples,

In this specific case, after knowing the scores, we can apply a *balance rule* to the PCA data, and make use of the *triangle rule*, as already reported in Chapter 2, in the mixture design section.

A Brief History of Principal Component Analysis

Beltrami (1873) and Jordan (1874) independently derived the SVD, and Fisher and Mackenzie in 1923 were the first to use it in a two-way analysis of an agricultural trial. It is generally accepted that the earliest descriptions of the PCA were given by Pearson (1901) and Hotelling (1933). Pearson in his paper comments about the computational aspect, and reports that his methods "can be easily applied to numerical problems", and also reporting that the calculations can become "cumbersome" for four or more variables. Hotelling started searching for a smaller "fundamental set of independent variables which determine the values of the original p variables". He was also the first to introduce the alternative term *components* to avoid confusion with the other uses of the word *factor* in mathematics. Other remarkable contributors are Girsch (30s), Rao (60s), Gower (60s), Jeffers (60s), and Presiendorfer and Mobley (80s). For a more detailed history of PCA, see (7).

3.3.6 CLUSTER ANALYSIS

We've seen in the previous sections that PCA can reveal groups in our data-set. Its primary aim indeed is *not* to give good separations between two groups or among groups. Other tools are better for this scope, and belong to the field of *cluster analysis*. In cluster analysis, the groups are *not known prior to the mathematical analysis.*

In order to individuate clusters we need to find objects *close* in the variable space and so the first priority will be to define a measurement of their *distance*.

Several definitions of distance can be employed, starting from the most common one- the *Euclidean distance*. If we have two objects in a multi-dimensional space of coordinates $(x1, x2, ..., xn)$ and $(y1, y2, ..., yn)$, their Euclidean distance will be defined by:

$$\sqrt{(x_1 - y_1)^2 + (x_2 - y_2)^2 + ... + (x_n - y_n)^n} \tag{3.3}$$

Considering again the data-set based on the work by Yucel (19), we can apply the Eq. 3.3 to calculate all distances between objects. We reported the standardized table data, while in the appendix, you can find how to perform the calculations also on the non-preprocessed data.

Distances can also be visually represented, as reported in Figure 3.23 for the autoscaled data, and in Figure 3.24 for the non-standardized data, where each square has a different intensity of color (or gray if reported in gray scale), depending on the value of the distance between objects.

After calculating the distances between the objects we can start searching for clusters in our data-set. What does it actually mean and how can we do it?

We can start from the easiest hypothesis of

- considering each object as a cluster, and then comparing its distance with all other objects

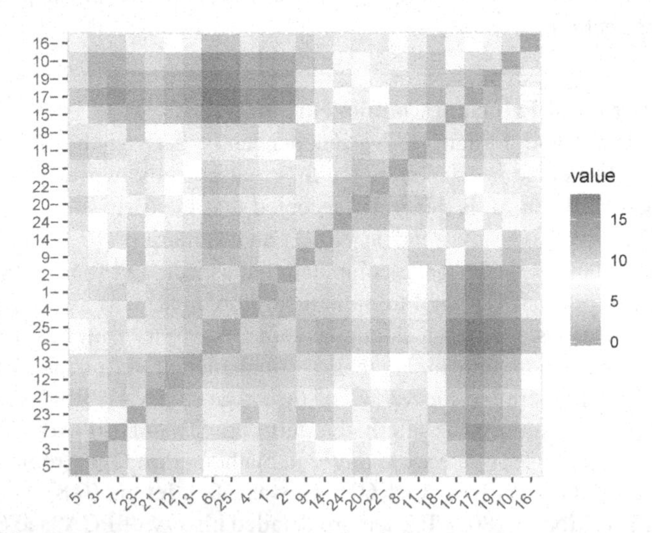

Figure 3.23 Plot of the distance matrix of the data-set based on (19). The data set is autoscaled. White colors represent a value of 6.

Table 3.6
Euclidean distances for the data-set based on the work by Yucel (19).

(table continued on the next page)

1	2	3	4	5	6	7	8	9	10	11	12
0.00	2.05	5.77	2.77	3.12	5.08	4.75	4.79	4.94	6.64	5.14	5.17
2.05	0.00	5.49	1.95	3.49	3.91	3.39	3.85	3.84	6.10	3.91	4.01
5.77	5.49	0.00	4.82	4.29	6.58	6.54	6.04	6.10	6.34	5.91	6.48
2.77	1.95	4.82	0.00	2.74	3.44	3.13	3.35	3.55	5.03	3.21	3.24
3.12	3.49	4.29	2.74	0.00	5.08	4.91	4.56	4.21	5.15	4.31	4.56
5.08	3.91	6.58	3.44	5.08	0.00	2.19	4.14	4.12	6.22	2.66	2.11
4.75	3.39	6.54	3.13	4.91	2.19	0.00	2.65	3.39	5.59	2.05	1.86
4.79	3.85	6.04	3.35	4.56	4.14	2.65	0.00	3.12	4.61	2.75	3.28
4.94	3.84	6.10	3.55	4.21	4.12	3.39	3.12	0.00	5.36	2.12	3.78
6.64	6.10	6.34	5.03	5.15	6.22	5.59	4.61	5.36	0.00	4.90	4.65
5.14	3.91	5.91	3.21	4.31	2.66	2.05	2.75	2.12	4.90	0.00	2.24
5.17	4.01	6.48	3.24	4.56	2.11	1.86	3.28	3.78	4.65	2.24	0.00
5.04	3.86	6.52	3.05	4.60	1.71	1.81	3.46	3.70	4.86	2.14	0.65
6.10	5.72	3.42	5.01	4.08	6.78	6.82	6.44	5.06	5.85	5.45	6.55
5.92	4.85	5.65	4.23	4.50	5.39	4.80	4.77	3.55	4.69	3.42	4.70
4.97	4.62	5.52	3.44	3.19	4.51	4.76	4.86	4.52	3.80	3.76	3.76
6.09	5.53	4.93	4.88	4.17	6.66	6.24	5.70	5.27	3.89	5.13	5.67
6.48	5.21	7.60	4.76	6.05	3.18	2.46	3.82	3.44	6.05	1.96	3.01
5.65	5.82	5.20	5.34	3.94	7.58	7.28	6.76	6.27	4.88	6.38	6.76
4.70	3.72	5.64	2.83	3.77	3.48	2.65	2.58	2.76	3.36	1.78	2.34
5.43	4.34	6.55	3.27	4.93	2.55	2.11	3.27	3.76	4.52	2.03	1.79
5.20	4.26	5.68	3.39	3.75	3.86	3.14	3.33	3.34	3.30	2.23	2.39
4.52	3.15	5.90	2.23	4.10	2.30	2.25	3.40	3.63	4.80	2.36	1.76
5.40	4.72	5.28	3.67	3.12	4.93	4.49	4.52	3.69	4.05	3.28	3.76
5.74	4.52	7.50	4.02	5.73	1.50	1.95	4.21	4.44	6.12	2.88	1.71

- consider the two closest objects as one cluster
- repeat the procedure of evaluating the distance among clusters

Of course if we repeat this process indefinitely we will obtain only *one* big cluster from our data-set. This is not our aim so we need to setup a *stop condition*.

Before doing that, we must also define a criteria for calculating the *distance between two clusters*, where one of the two (or both) contain more than one objects. One of the easiest ways is to use the table of the distances and select the one referred to the nearest neighbors (and repeat the process for all the clusters involved).

In order to help the reader in understanding the process, we will perform these steps on a small subset of the data-set shown in Table 3.7 and 3.3.6, where we refer to the concentrations of Ag and Fe.

13	14	15	16	17	18	19	20	21	22	23	24	25
5.04	6.10	5.92	4.97	6.09	6.48	5.65	4.70	5.43	5.20	4.52	5.40	5.74
3.86	5.72	4.85	4.62	5.53	5.21	5.82	3.72	4.34	4.26	3.15	4.72	4.52
6.52	3.42	5.65	5.52	4.93	7.60	5.20	5.64	6.55	5.68	5.90	5.28	7.50
3.05	5.01	4.23	3.44	4.88	4.76	5.34	2.83	3.27	3.39	2.23	3.67	4.02
4.60	4.08	4.50	3.19	4.17	6.05	3.94	3.77	4.93	3.75	4.10	3.12	5.73
1.71	6.78	5.39	4.51	6.66	3.18	7.58	3.48	2.55	3.86	2.30	4.93	1.50
1.81	6.82	4.80	4.76	6.24	2.46	7.28	2.65	2.11	3.14	2.25	4.49	1.95
3.46	6.44	4.77	4.86	5.70	3.82	6.76	2.58	3.27	3.33	3.40	4.52	4.21
3.70	5.06	3.55	4.52	5.27	3.44	6.27	2.76	3.76	3.34	3.63	3.69	4.44
4.86	5.85	4.69	3.80	3.89	6.05	4.88	3.36	4.52	3.30	4.80	4.05	6.12
2.14	5.45	3.42	3.76	5.13	1.96	6.38	1.78	2.03	2.23	2.36	3.28	2.88
0.65	6.55	4.70	3.76	5.67	3.01	6.76	2.34	1.79	2.39	1.76	3.76	1.71
0.00	6.46	4.59	3.67	5.73	2.86	6.78	2.35	1.51	2.57	1.49	3.85	1.41
6.46	0.00	3.91	4.52	3.50	7.10	3.80	5.13	6.38	5.10	5.84	4.06	7.58
4.59	3.91	0.00	3.44	2.82	4.49	4.42	3.04	4.16	2.91	3.93	2.52	5.63
3.67	4.52	3.44	0.00	3.25	5.12	3.92	2.83	3.63	2.60	3.21	2.50	4.92
5.73	3.50	2.82	3.25	0.00	6.56	2.34	4.06	5.55	3.65	5.02	3.04	7.04
2.86	7.10	4.49	5.12	6.56	0.00	7.86	3.13	2.44	3.43	3.50	4.66	2.77
6.78	3.80	4.42	3.92	2.34	7.86	0.00	5.17	6.68	4.91	6.15	4.14	8.06
2.35	5.13	3.04	2.83	4.06	3.13	5.17	0.00	1.96	1.25	2.35	2.73	3.54
1.51	6.38	4.16	3.63	5.55	2.44	6.68	1.96	0.00	2.45	1.99	3.83	2.21
2.57	5.10	2.91	2.60	3.65	3.43	4.91	1.25	2.45	0.00	2.62	1.99	3.79
1.49	5.84	3.93	3.21	5.02	3.50	6.15	2.35	1.99	2.62	0.00	3.52	2.52
3.85	4.06	2.52	2.50	3.04	4.66	4.14	2.73	3.83	1.99	3.52	0.00	5.06
1.41	7.58	5.63	4.92	7.04	2.77	8.06	3.54	2.21	3.79	2.52	5.06	0.00

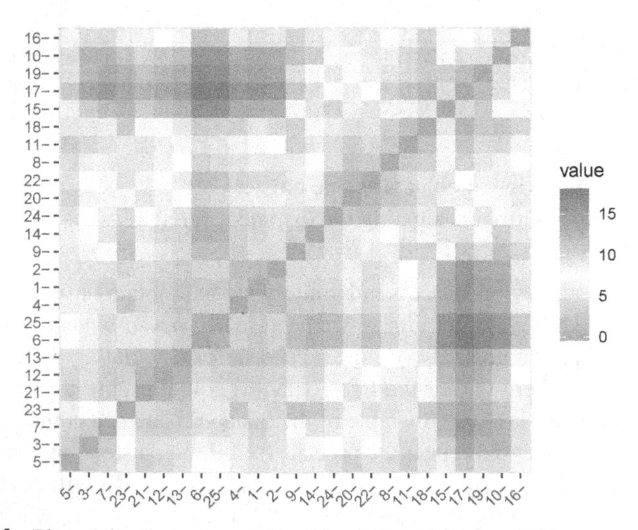

Figure 3.24 Plot of the distance matrix of the data-set based on (19).

Table 3.7
Data used for clustering the first five points of the data-set based on (19).

Ag	Fe
-0.02	0.15
-1.55	-0.87
-0.40	-0.95
0.70	-1.22
1.03	1.42

1	2	3	4	5
0.00	1.84	1.17	1.55	1.65
1.84	0.00	1.15	2.28	3.45
1.17	1.15	0.00	1.14	2.77
1.55	2.28	1.14	0.00	2.66
1.65	3.45	2.77	2.66	0.00

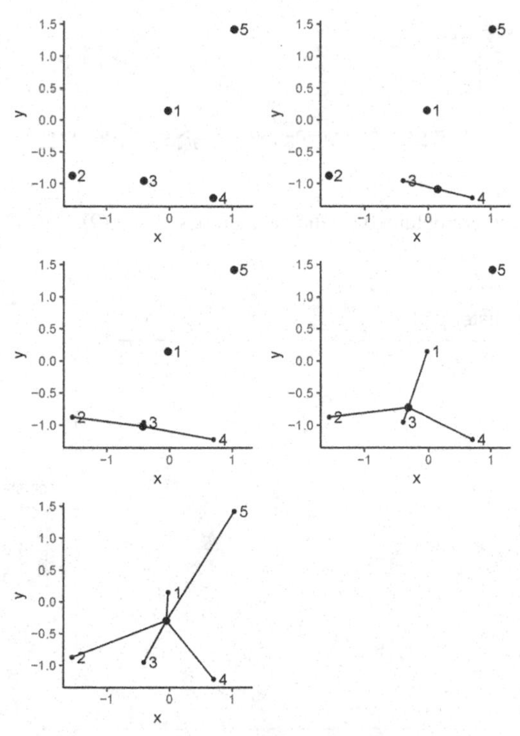

Figure 3.25 Process of clustering. In the first step, each object is considered a cluster (top left). In the second step, the two nearest objects form the first cluster (top right). Starting from the second step, each object is added to the cluster using the rule of the nearest neighbors. x and y refer to the concentrations of Ag and Fe ions, as reported in 3.7.

3.3.7 DENDROGRAMS

Another way of calculating the composition of the groups for the data-set in exam is to resort to a parameter defined as *similarity* using the following formula:

$$s_{ij} = 100(1 - d_{ij}/dmax)$$

where i and j represent the index of two points. The stop condition will be determined by the user.

If we report the *distances* or the *similarities* in a plot we obtain a *dendrogram*.

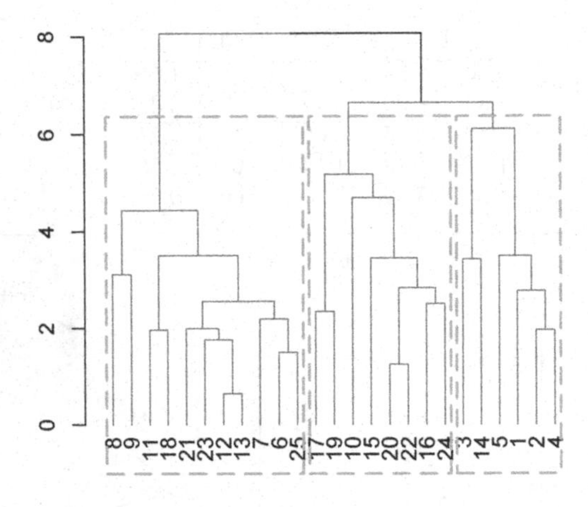

Figure 3.26 Dendrogram obtained on the data-set based on (19).

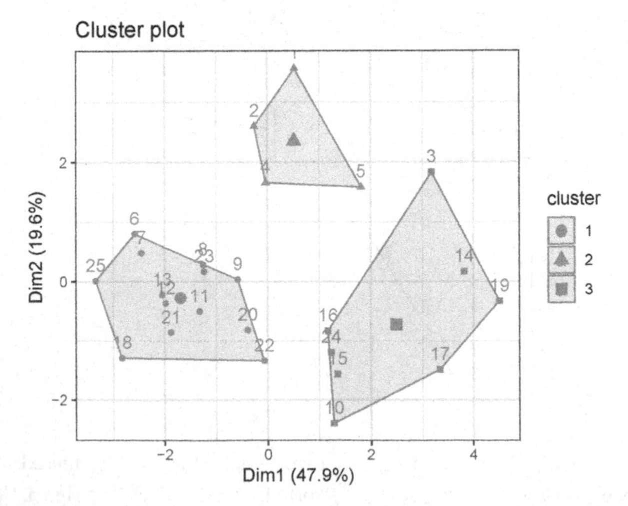

Figure 3.27 K-means calculated on data-set based on (19), considering three clusters as an initial starting condition. Data-set was standardized before the analysis.

The clustering just described is a *hierarchical method* and it means that when an object is assigned to a group this process cannot be reversed. The opposite case is called *non-hierarchical.*

3.3.8 K-MEANS METHOD

An example of the *non-hierarchical* method is the *k-means means.* The steps for applying it to a data-set are the following:

- divide the points into k clusters
- assign each individual to the cluster whose centroid is nearest
- if the cluster considered in the previous step loses or gains one individual, recalculate the position of its centroid
- reiterate the process and stop when all the points are in the cluster whose centroid is nearest

An alternative way is also to consider *seed points* instead of clusters. We now apply this method to the data-set based on (19) after scaling and considering three clusters.

The first problem you will notice is a lack of consistency if we repeat the analysis. Why? Because the method involves an initial arbitrary grouping of the samples that will reflect on the results. So repeating the analysis with a different grouping will give different results. The second problem is related to the number of clusters. What would have happened with a different initial condition for the number of clusters?

The algorithm would have performed the clustering based on our initial choice without any problem. So how do we choose the initial number of clusters? There

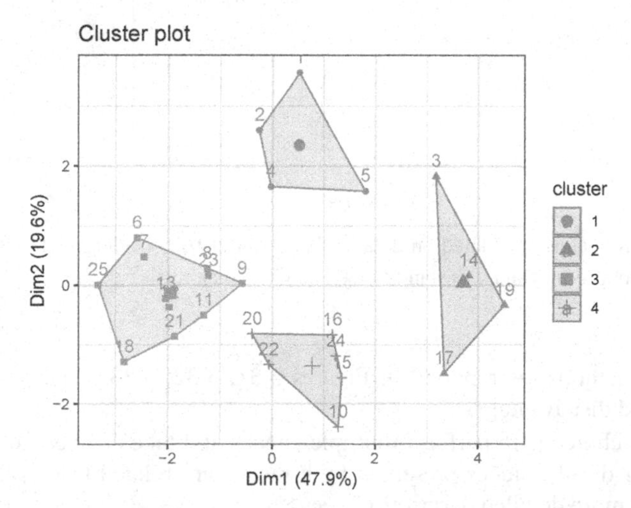

Figure 3.28 K-means calculated on data-set based on (19), considering four clusters as an initial starting condition. Data was standardized before the analysis.

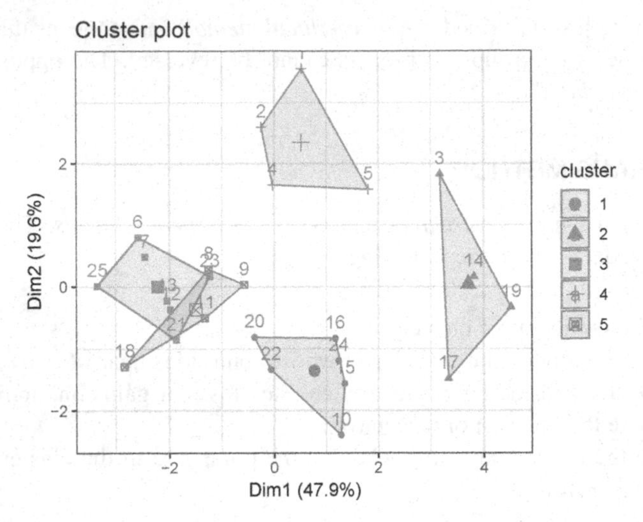

Figure 3.29 K-means calculated on data-set based on (19), considering five clusters as an initial starting condition. Data was standardized before the analysis.

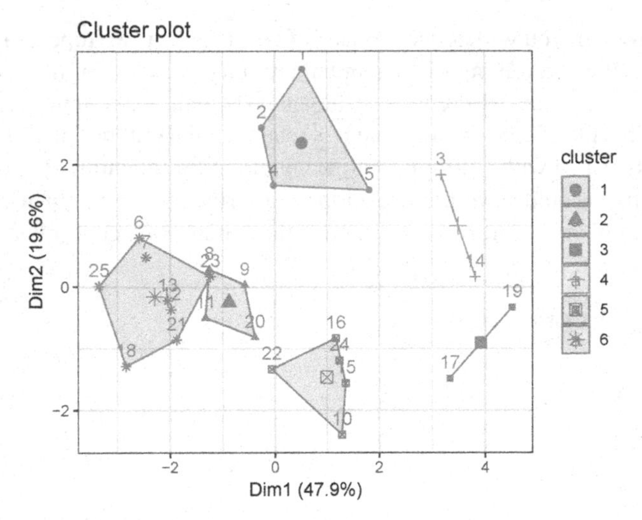

Figure 3.30 K-means calculated on data-set based on (19), considering six clusters as an initial starting condition. Data was standardized before the analysis.

are different methods (as reported in Figures 3.31, 3.32, 3.33), each with its own advantages and disadvantages.

A practical choice is to perform multiple checks and take an average number of clusters. Other disadvantages presented by k-means are related to its sensitivity to outliers. For a more detailed description, see (8).

For our data-set, two out of three tests give an optimal number of clusters of three.

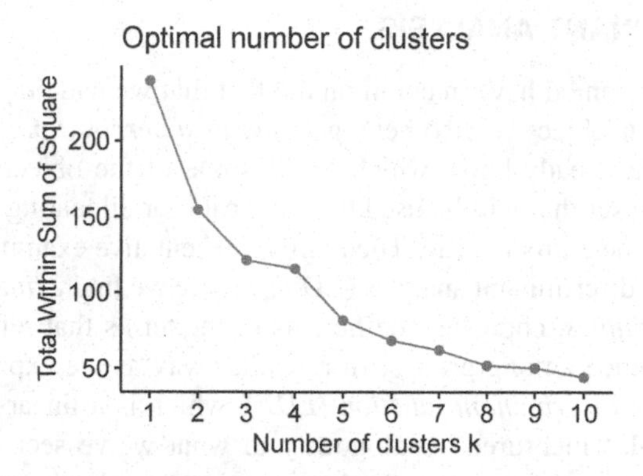

Figure 3.31 Diagnostics (Total within Sum of Square) on K-means calculated on data-set based on (19).

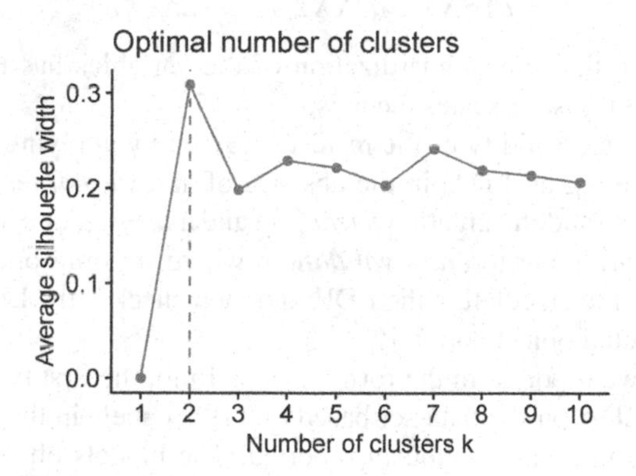

Figure 3.32 Diagnostics (Average silhouette width) on K-means calculated on data-set based on (19).

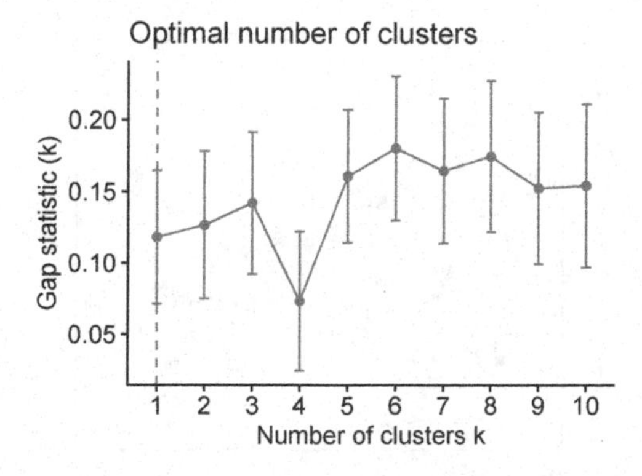

Figure 3.33 Diagnostics (Gap statistic) on K-means calculated on data-set based on (19).

3.3.9 DISCRIMINANT ANALYSIS

All the methods presented have in common the fact that we had *no prior knowledge* of the groups of our object, and so belong to the *unsupervised pattern recognition methods*. When we already know which group some of the objects belong to, we can build a training set that will be used to find a rule for allocating a new object of unknown group to the correct group. The more representative example of these techniques is the linear discriminant analysis (LDA), where we find a *linear combination of the original variables*, choosing coefficients of the terms that reflect *as much as possible* the difference among groups. In a similar way as we expressed PCA, we will obtain a *linear discriminant function (LDF)*, which is a linear combination of the original variables measured. Analogously to what we've seen in the previous visual representation, we can report the LDF in a scatter plot in order to visualize clusters in our data.

$$Y = a1XX1 + a2XX2 + \ldots + anXXn$$

It's due to notice that the standardization of the variables has no effect on the results of LDA, and it just re-scales the axis.

How can we test the validity of the results obtained by applying a LDA classification? Before applying an LDA, in the absence of an *external* test set, we should divide our data-set at random, creating a *training* and a *test set*, or eventually (but not as preferred solution), recur to *cross-validation*, where we omit one (or a small selection of) object while calculating the LDF, and then check if the LDF just obtained can predict the omitted object correctly.

In Figure 3.34, we reported in the form of scatter plot the first two LDF obtained by performing an LDA on the data-set based on (19). Labels in the clusters refer to the origin of the sample, as previously reported. The models obtain the following accuracy in prediction- class a = 100%, M = 87.5%, and T = 100%, with a total accuracy percentage of 96%. In order to calculate the model, we scaled the data

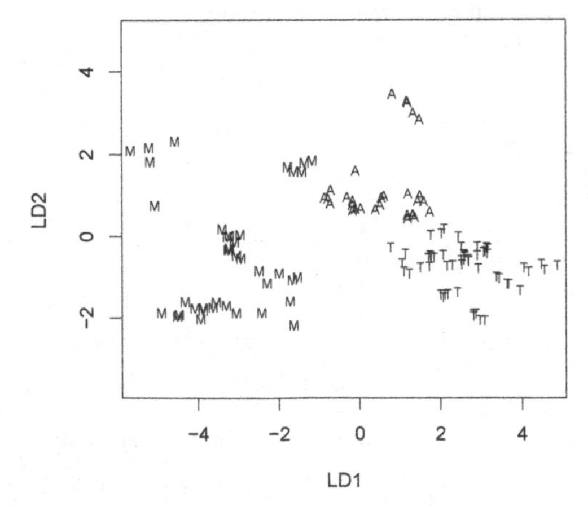

Figure 3.34 LDA analysis performed on data-set based on (19).

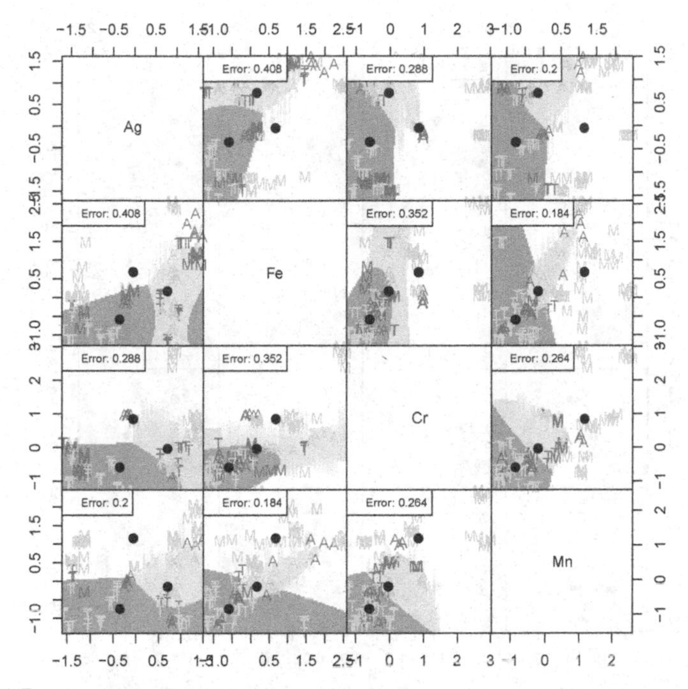

Figure 3.35 Quadratic discriminant analysis performed on (19). For elements Ag, Fe, Cr and Mn.

and partitioned the data in a training and test set (80% and 20% of the samples, respectively). The readers are suggested to repeat the calculations with the help of the script presented in the appendix with and without scaling, and also by varying the partitioning of the samples.

Is it possible to enhance the performance of the model in order to obtain better prediction results? Even if in the specific case reported there is probably no need to do it, after applying a Linear Discriminant Analysis, we can try to model using *quadratic functions* and so perform a *quadratic discriminant analysis*. A visual representation is visible in Figures 3.35, 3.36, and 3.37, where in each tile, the classification order and the apparent error rates are given.

The performance increased as expected obtaining a 100& accuracy for all classes.

We finally remark that all the techniques presented, instead of the original variables, can make use of the *linear combinations of the variables* before the discriminant analysis (i.e., PCs can also be used in this scope) to reduce the information of the data-set. The reader is also advised to try this approach on the data-set, and compare the results.

3.3.10 SOFT INDEPENDENT MODELLING OF CLASS ANALOGY

So far, we presented modelling techniques where the final aim was to obtain the best description among groups, while in some occasions, we prefer to know whether or not an object belongs to one specific class. For this purpose, we can find a *separate*

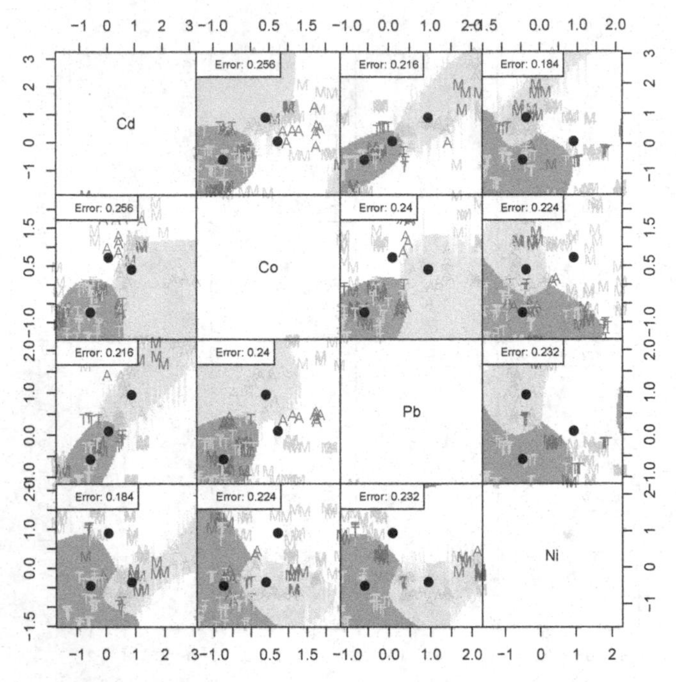

Figure 3.36 Quadratic discriminant analysis performed on (19). For elements Cd, Co, Pb, Ni.

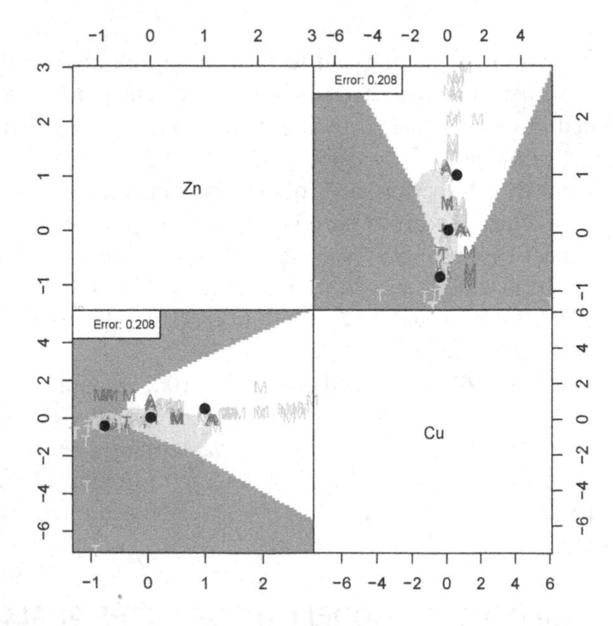

Figure 3.37 Quadratic discriminant analysis performed on (19). For Elements Zn and Cu.

model for each class, and then *test whether or not the object is a member of it*. SIMCA belongs to this kind of methodology, allowing the user to create a model for each class, making use of the principal component.
Again the performance of this methodology can be reported by means of a visual representation of the model, performances of the model, and its prediction abilities. Applying the methodology to the data-set previously used for explaining LDA, we choose to present the following plots:

The model plots report scores and loading plot (top row), a distance plot score (T2, h) vs orthogonal (Q, q) distances, and corresponding critical limits for given number of components, and a plot of the cumulative variance. The predictions plots report sensitivity and ratio of misclassified values, depending on the number of components. A prediction plot is also reported, which shows classification results for each object. Finally, we report a *Cooman's plot*. They show an orthogonal distance, q, from objects to two selected classes/models. In figures, we reported for pairs A,T and M,T. The closest the points are in the plot, the "more similar" the objects are. Having a look at the values and plot, SIMCA modelled almost flawless the class present in the data-set, so the model looks promising for being tested with an external test set.

We can summarize the results for each group:

Origin *Aegean* (A):

- Distances plot presents only one value which is suspicious (outside the 95% confidence interval)

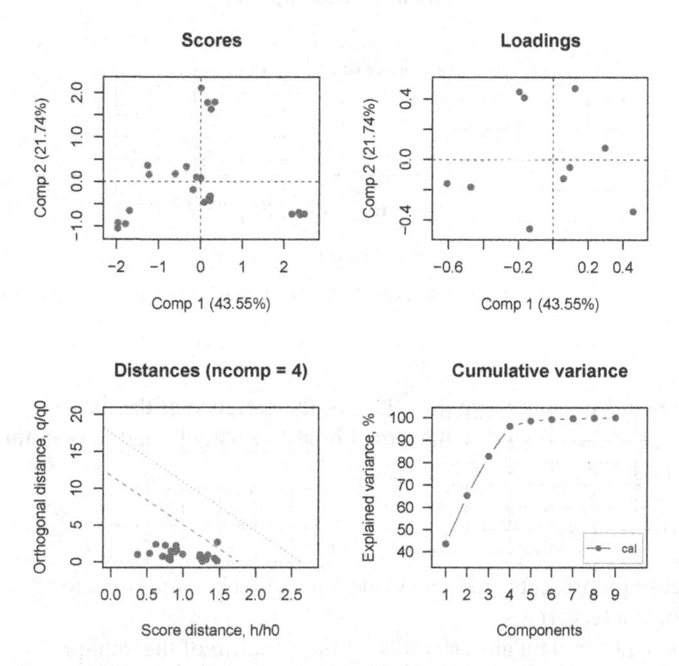

Figure 3.38 SIMCA analysis performed on (19). Model plot for class Aegean (A).

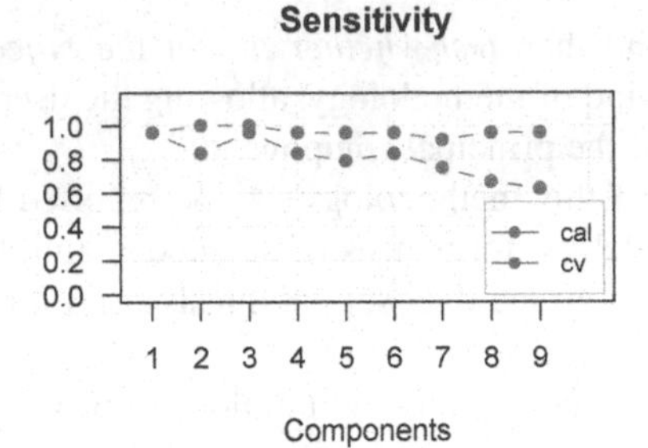

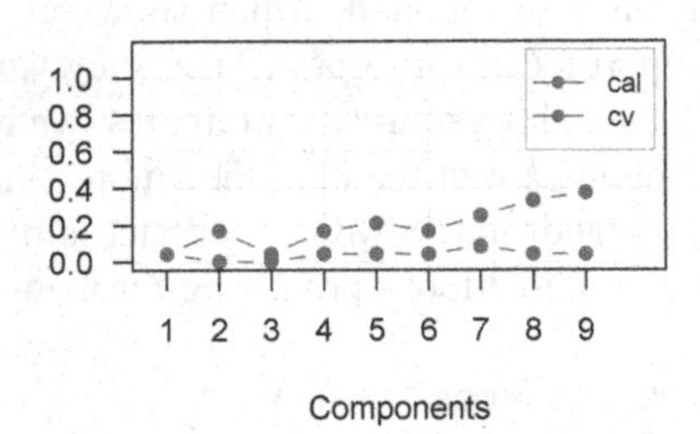

Figure 3.39 SIMCA analysis performed on (19). Sensitivity and misclassified plot for class Aegean (A).

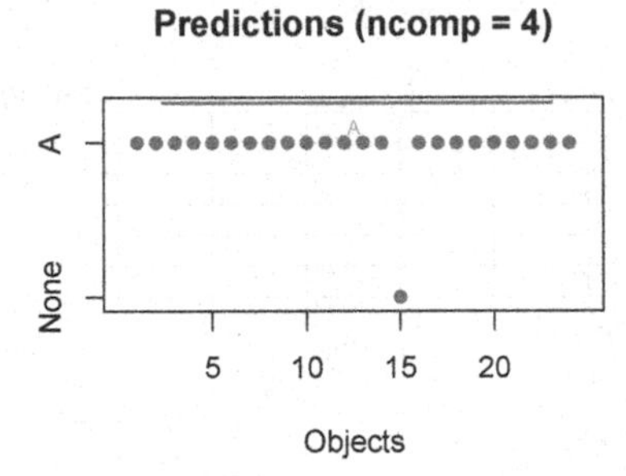

Figure 3.40 SIMCA analysis performed on (19). Sensitivity and misclassified plot for class Aegean (A).

- 4 components explain more than 95% of the variance of the data-set
- a model with 3 components have the highest sensitivity and lowest number of misclassified objects

Origin *Marmara* (M):

- Distances plot presents only one value which is suspicious (outside the 95% confidence interval)
- 6 components explain almost 100% of the variance of the data-set
- a model with 2 components has the highest sensitivity and lowest number of misclassified objects

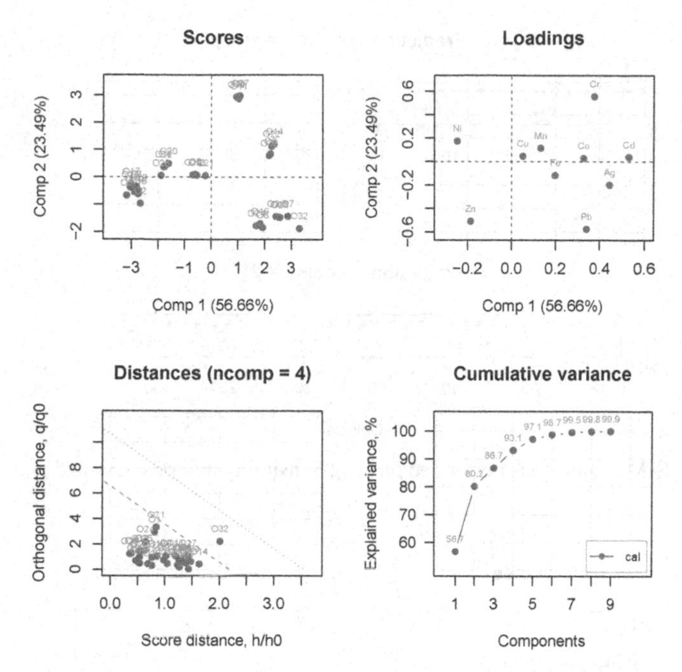

Figure 3.41 SIMCA analysis performed on (19). Model plot for class Marmara (M).

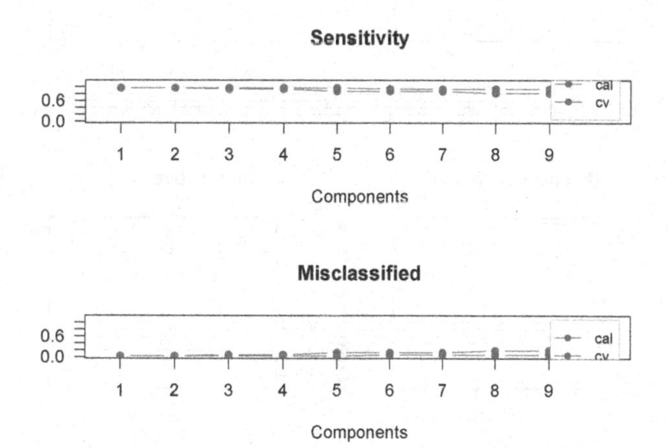

Figure 3.42 SIMCA analysis performed on (19). Sensitivity and misclassified plot for class Marmara (M).

- selecting 2 or 4 components does not change the number of misclassified objects.

Origin *Thrace* (T):

- Distances plot presents at least 3 values near the 95% confidence interval
- 6 components explain more than 95% of the variance of the data-set

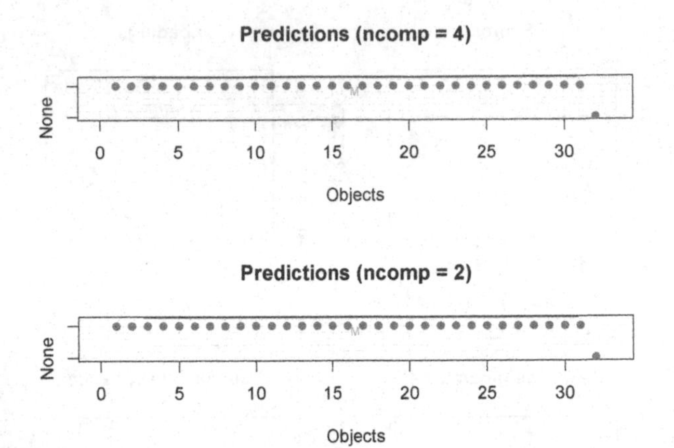

Figure 3.43 SIMCA analysis performed on (19). Sensitivity and misclassified plot for class Marmara (M).

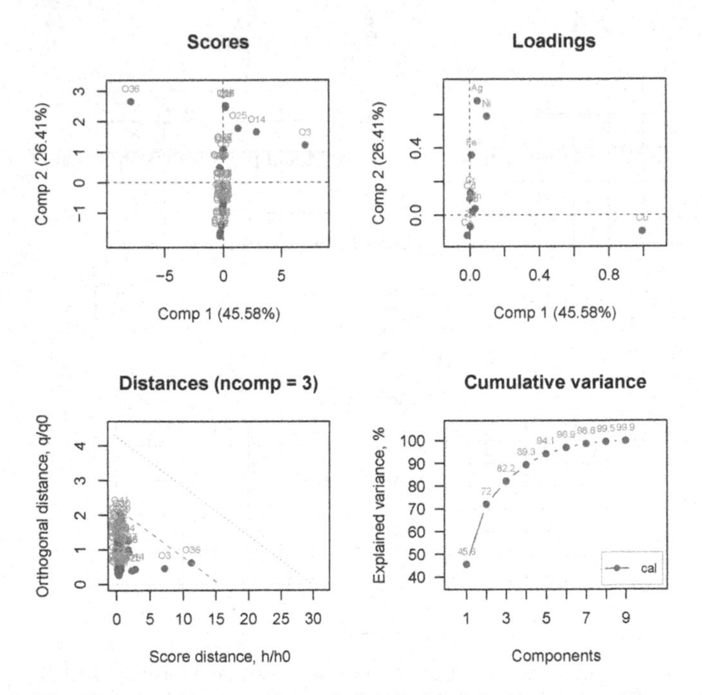

Figure 3.44 SIMCA analysis performed on (19). Model plot for class Thrace (T).

- a model with 2 components has the highest sensitivity and lowest number of misclassified objects
- selecting 2 or more components does not change the number of misclassified objects.

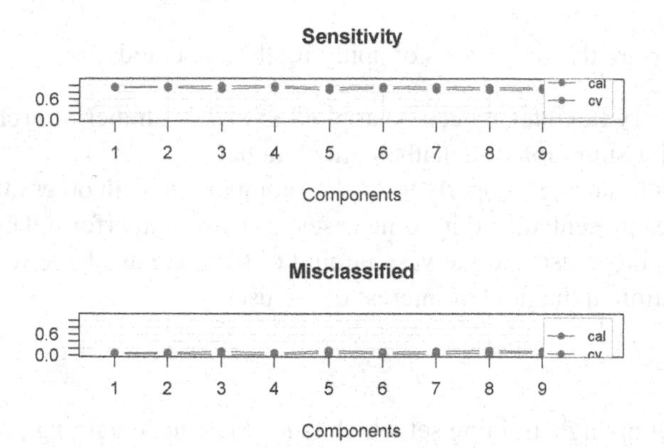

Figure 3.45 SIMCA analysis performed on (19). Sensitivity and misclassified plot for class Thrace (T).

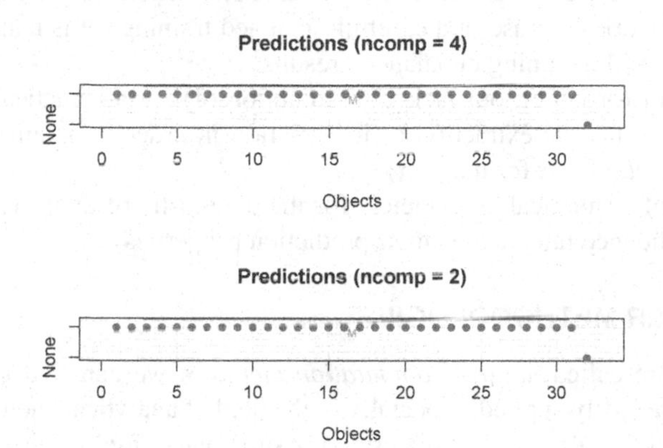

Figure 3.46 SIMCA analysis performed on (19). Sensitivity and misclassified plot for class Thrace (T).

3.3.11 ARTIFICIAL NEURAL NETWORKS

A very hot topic, as testified by the vast quantity of literature present, is the adoption of Artificial Neural Networks (ANN) in almost every field. Materials science has embraced these trends with fruitful results. What exactly are ANN?

ANN, inspired by our knowledge of how the brain works, receive our input data in what is called an *input layer*, and then process by an intermediate layer (the so-called *hidden layer*), to produce one or more output. As a practical example, we can input to our network the absorbance spectra in order to obtain the concentration of a specific analyze as output. The hidden layer in this particular case should be trained to give the correct weight in order to find how we can obtain the final concentration of the solutions from the starting spectra. To do so, we need to *train* the network using a *training set* (and also in this case, the bigger the better). The model, or better the *network*, as in the previous methodologies reported, will be tested by means of a *test set*.

What exactly are the advantages of applying these methods?

- they are very flexible and can be applied to a vast number of problems due to lack of assumption of a mathematical model
- their performance, if properly trained, is comparable with other multivariate techniques presented, and in some cases, can also outperform them
- their ubiquitous use and the vast amount of software available for their implementation in the field of interest of the user

What are the caveats?

- the risk of giving a training set which is *too specific*, obtaining poor performance in the prediction, in other words *overfitting*
- the lack of a signifying training set. The number of samples to be used to train a network depends on several parameters, including the architecture of the network. The use of a carefully planned training set is mandatory in order to avoid obtaining unbalanced results
- the lack of (or better, our lack of assumption of) a mathematical model a priori, prevents the extraction of information in order to obtain one (they work as a *black box* for the user)
- the lack of "canonical" parameters for the diagnostic of their performance (i.e., confidence intervals on their prediction properties)

3.3.12 OTHER METHODOLOGIES

Among what are called *natural computation methods*, we can find *generic algorithms*. They are vastly applied, especially in the field of analytical chemistry (and so in the field of characterization of materials). As the name implies, they mimic their biological background. We can again use as an example a data-set of spectra and the concentration of a specific analyte. At first, a selection of frequencies will be performed (an *initial population*), and based on this selection, we will test how we can predict the concentration of our solution (i.e., using multivariate techniques already presented). Then we will combine or select the original variables, applying processes that mimic *inheritance*, selection of *suitable parents*, *crossover*, and *mutation*.

3.3.13 Q.A.

Q. Should I always *autoscale* my data?

A. Let's say we built a data-set where we already know that the scale of one of the variables is very different from the other variables. If we will compare its influence with the other factors, it is quite intuitive that it will be weighed by its value while performing a PCA. To avoid this we can *normalize* the data in order that all the values of the variable will lie in similar ranges. One of the most common methodologies is to normalize the data according to their mean and variance. This is only one of the several preprocessing techniques we can use. Another common technique, due to the nature of the data gathered in material science, is also to apply non-linear methods based on functions that can change the range of our data in a specified interval. One of the problems of applying these kinds of multivariate methods, considering them as blackboxes, is to use the default values for the software settings, or blindly applying the latest setup used. Unfortunately, this can be quite misleading if not avoided.

Q. Is it better to use the data-set as is, or is it better to group the samples according to the replicates of the samples?

A. A lot of software will perform the averaging for you. So, at least at first, I would perform the data considering all replicates in order to also spot the outliers in the data-set.

Q. Do I need to perform any test to check the distribution of my data-set before performing a PCA?

A. PCA needs no distributional assumption and as such is an adaptive exploratory method which can be used on numerical data of various origins.

Q. Is principal component analysis the same as factor analysis?

A. As it can be found in (7), factor analysis is rather different from principal component analysis. Both the techniques are aimed at reducing the dimensionality of a data-set. Factor analysis assume a definite model, while PCA assumes no factor analysis, and PCA tries to represent some aspect of the covariance matrix Σ (or correlation matrix). To further clarify the difference between principal component analysis and factor analysis, we can also refer to Mukhopadhyay (12): *In principal component analysis, principal components are linear functions of the variables, while in factor analysis, the variables are expressed as linear combinations of factors*. And also, *in component analysis the emphasis is on the explaining the total variance* $\sum_{i=1}^{p} s_{ii}$, *while factor analysis attempts to explain the full covariance matrix S*.

Q. What is orthogonality and why is it important?

A. Orthogonality is important in statistics, partly because orthogonal vectors often correspond to random variables with zero correlation. Many procedures in estimation and analysis of variance are nothing more than the decomposition of the data vector into orthogonal pieces.

Q. Can I use PCA also if I know that my variables are not "strongly" correlated?

A. PCA is a useful technique for reducing the amount of data only if there is a correlation present. It is not a useful technique if the variables are uncorrelated. Also, outliers have a severe influence on the results. In order to avoid this problem, since outliers increase the variance in an uninformative direction, we can make use of *robust pca* methodologies that determine the PCA directions to maximize a robust measure of variance.

Q. Is it always necessary to use PCs before applying an LDA?

A. This is an interesting question, and I have to say that I did not ask it myself before stumbling via serendipity to the very instructional tutorial written by Peter Nistrup "https://towardsdatascience.com/linear-discriminant-analysislda-101-using-r-6a97217a55a6". In order to check, I re-performed the analysis also on our data-sets, obtaining the following results. I guess that I'm quite convinced from now on to apply LDA on the original data, and to perform a quick check to see if the PC version can give better results, but due to the advantage of using the original space, I will stick to the original variables for the interpretation of the results obtained.

3.3.14 EXERCISES

The readers can use the data-sets included performing experiment and:

- perform an exploratory principal component analysis *selecting* the suitable preprocessing
- eventually compare the results without applying any preprocessing
- apply non-hierarchical classification techniques and compare the results with the workflow reported in the appendix

3.3.15 REMARKS

One of the problems I first encountered while applying multivariate techniques to my workflow was the application of principal component analysis to impedance measurements. Why? This is because, as probably the readers know, the results obtained from these techniques are in the form of *complex numbers*. Since complex number have their own properties, the question was if the methodologies of chemistry can be applied as is, or if they need to be changed, and most of all, if the interpretation changes.

The short answer is that indeed we can apply chemometric methodologies as the principal component analysis; just that we need to make some changes. The problem is that there is a scarcity of literature and software ready to deal with these issues. The reader is suggested to refer to the pioneering work of Geladi for more details (4).

3.3.16 SUGGESTED ESSENTIAL LITERATURE

An essential guide for pattern recognition is the book *Pattern Recognition* by Sergios Theodoridis (15). It is exhaustive. It is also more versed on a theoretical approach (from the point of view of a chemist), and this can be an advantage or disadvantage depending on the background of the reader.

A comprehensive text with a more practical approach, with a wide selection of case studies is "Chemometrics for pattern recognition" by Richard Brereton (2).

The book by Miller (11) offers a concise and practical approach in a clear manner, which is a great starter for practitioners, and for sure is worth reading. A comprehensive book that can be used as a reference for interpreting biplot is *Understanding Biplots* by Gower et al. (6).

For performing pattern classification using the software R, I suggest the reader to rely on the textbook by Kassambara (8), and the corresponding web resources, the book written by Schumaker *Using R with multivariate statistics* (14), and also the text from Wehrens *Chemometrics with R.* (18).

Bibliography

1. F. J. Anscombe. Graphs in statistical analysis. *The American Statistician*, 27(1):17–21, 1973.
2. Richard G. Brereton. *Chemometrics for Pattern Recognition*. Wiley, 2009.
3. Sangit Chatterjee and Aykut Firat. Generating data with identical statistics but dissimilar graphics: A follow up to the anscombe dataset. *American Statistician*, 61(3):248–254, aug 2007.
4. Paul Geladi, Andrew Nelson, and Britta Lindholm-Sethson. Complex numbers in chemometrics. Examples from multivariate impedance measurements on lipid monolayers. *Analytica Chimica Acta*, 2007.
5. Mengjing Hou, Yu Huang, Lingwei Ma, and Zhengjun Zhang. Compositional Analysis of Ternary and Binary Chemical Mixtures by Surface-Enhanced Raman Scattering at Trace Levels. *Nanoscale Research Letters*, 10(1):1–7, dec 2015.
6. Niel le Roux John Gower Sugnet Lubbe. *Understanding Biplots*. John Wiley & Sons, 2010.
7. I.T. T Jolliffe. *Principal Component Analysis, Second Edition, New York: Springer Science & Business Media*. Springer Nature, 2002.
8. Alboukadel Kassambara. *Practical Guide to Cluster Analysis in R, Unsupervised Machine Learning*. CreateSpace Independent Publishing Platform, 2020.
9. F.P. La Mantia, L. Botta, M. Morreale, and R. Scaffaro. Effect of small amounts of poly(lactic acid) on the recycling of poly(ethylene terephthalate) bottles. *Polymer Degradation and Stability*, 97(1):21–24, jan 2012.
10. Justin Matejka and George Fitzmaurice. Same stats, different graphs: Generating datasets with varied appearance and identical statistics through simulated annealing. In *Conference on Human Factors in Computing Systems - Proceedings*, volume 2017-May, pages 1290–1294. Association for Computing Machinery, may 2017.
11. N.Miller Miller and Jane C. Miller. *Statistics and Chemometrics for Analytical Chemistry*. Prentice Hall, 2010.
12. Parimal Mukhopadhyay. *Multivariate Statistical Analysis*. World Scientific, 2008.
13. Ruibo Ren, Kechao Han, Pinhui Zhao, Jingtao Shi, Lei Zhao, Dongxing Gao, Zeyu Zhang, and Ziqiao Yang. Identification of asphalt fingerprints based on ATR-FTIR spectroscopy and principal component-linear discriminant analysis. *Construction and Building Materials*, 198:662–668, feb 2019.
14. Randal E. Schumaker. *Using R With Multivariate Statistics*. SAGE Publications, Inc, 2015.
15. Sergios Theodoridis and Konstantinos Koutroumbas. *Pattern Recognition (Fourth Edition)*. Academic Press, 2009.
16. Timothy C. Urdan. *Statistics in plain English*. Lawrence Erlbaum Association, 2005.

17. Kurt Varmuza and Peter Filzmoser. *Introduction to Multivariate Statistical Analysis in Chemometrics*. CRC Press, 2016.
18. Ron Wehrens. *Chemometrics with R*. Springer, 2011.
19. Yasin Yücel and Cevdet Demir. Principal component analysis and cluster analysis for the characterisation of marbles by capillary electrophoresis. *Talanta*, 2004.

4 Calibration

Objectives:

- learn basic concepts of univariate and multivariate calibration
- know the conditions for applying a multivariate regression to a data-set
- understand the basic principles of principal component regression
- understand the basic principles of partial least square regression

4.1 INTRODUCTION

Calibration involves using the *data gathered from our system* by means of diagnostic techniques and using them to *predict one or more of its property or underlying parameter(s).*

It is widely studied and employed in the field of chemometrics and in the field of analytical chemistry since its beginnings.

Due to the wide use of analytical techniques in Material Science, it has been widely applied to predict the mechanical properties of alloys, composite systems, and polymers. It also finds applications in the properties of systems (e.g., corrosion studies), and also in the "traditional" field of determining the concentration of a specific analyte.

In this chapter, we will describe calibration techniques related to univariate and multiregression methods, focusing on principal component regression, partial least squares regression, and finally, we will present a general overview of other multivariate techniques.

4.2 UNIVARIATE CALIBRATION

The first regression performed by a chemist has probably involved the application of Lambert Beer Law.

One of the first tutorial experiment performed in an analytical laboratory is to find the concentration of an unknown analyte measuring a series of spectra, selecting one specific wavelength and then measuring the value of Absorbance (or Transmission) at that specific wavelength. Repeating the measurement with different solutions at different concentrations, let the user create a data-set of Absorbance (or Transmission) vs Concentration that can be reported in the form of a scatter plot, and also let us calculate a linear interpolation of the results gathered.

I would like also to remind the reader that

- Since we *already* know that there is a *linear* correlation between the concentration and the absorbance, the best experimental points to measure the spectra are the lower, higher, and half concentration, as described by the design of the experiment.
- We interpolate using a linear model because we *already* know the laws followed by our system. If these are unknown, we need to rely on another experimental plan.

As a practical example, if we are allowed to perform ten experiments, the better way to distribute them in the experimental domain is to perform four measurements at lower concentration, four at higher concentration, and finally two measurements at 50% concentration.

Distributing the concentration uniformly inside the interval of investigation (i.e., a measurement every 10% of concentration) is useful *ONLY* if we need to check if the relationship between concentration and our response is linear. If we want to write down the equations related to the process presented, we can write

$$x = a_0 + a_1 y + error$$

where (a_0 and a_1 are two constant values).

We can then fit this model using least squares, inverting the equation to give

$$y = -(a_0/a_1) + (1/a_1)x$$

and then use the model obtained for prediction. This is the *classical* way for a chemist to report a calibration and, as reported by Naes (3), referred to as *classical calibration*.

Another approach, introduced lately in statistics, is to perform the opposite process, which is then called *inverse calibration*. In this case the model can be written as

$$y = b_0 + b_1 x + error$$

then fit using least squares and used in the form

$$y = b_0 + b1x$$

for prediction.

What is the difference between the two approaches? Summarizing, if the fit is good using the first of second approach should negligible. This answer implies that we need to have a parameter to define a *good fit*. The first parameter is the R^2, where R is the *correlation coefficient* that measures how well the prediction agrees with the references (values between -1 and 1), while the other one is the mean square error (MSE) that is *sum of the squared difference between the predicted values and the real values*. In general terms, inverse methods predict better values close to the mean, and work better if the calibration samples represent a random selection from a population. For more details about the topic the reader is suggested to see (3).

4.3 UNIVARIATE CALIBRATION, DATA-SET CONCRETE

The application of linear modelling is not limited to spectroscopy and can be employed in a wide range of applications. The following example of calibration is based on the work of (8), where they studied the influence of carbon nanotubes (CNTs) on the compressive strength and air-void structure of ultra-light foamed concrete. As it can be read in their paper, the authors, in order to obtain an enhancement of the properties of ultra-light foamed concrete and in particular to increase the strength, added CNR to traditional foamed concrete. They prepared foamed concrete with a density between 200 kg/m^3, 250 kg/m^3, with a variable content from 0.05% to 0.35% of CNTs, and then measured the compressive strength and the porosity and air-void size. Data-sets are reported in Table 4.1.

Due to the linear relationship present between concrete strength and porosity, we can perform a regression using the following generic formula:

$$y = \alpha + \beta \cdot x.$$

In Figures 4.1 and 4.2, we report the plot of porosity (%) vs compressive strength (MPa), and air-void size (μm) vs compressive strength (MPa), including the linear models calculated (continuous line).

For both the models calculated, there is a good fitting relation, as also confirmed by the very low p-value reported in Table 4.2. The parameters for the regression are reported in Figure 4.3. The value of Adjusted R-squared is 0.9957 for the model of compressing strength dependent on porosity, while it is 0.9125 for the model of

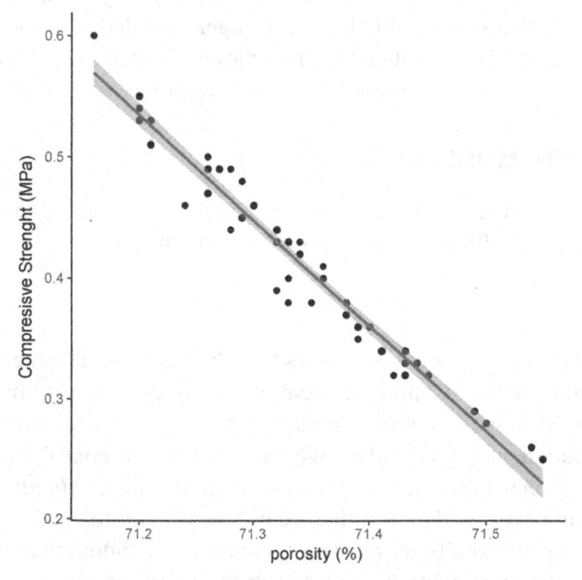

Figure 4.1 Linear regression for porosity (%) vs compressive strength (MPa) for the data-set based on (8). The grayed areas represent a 95% confidence interval in prediction.

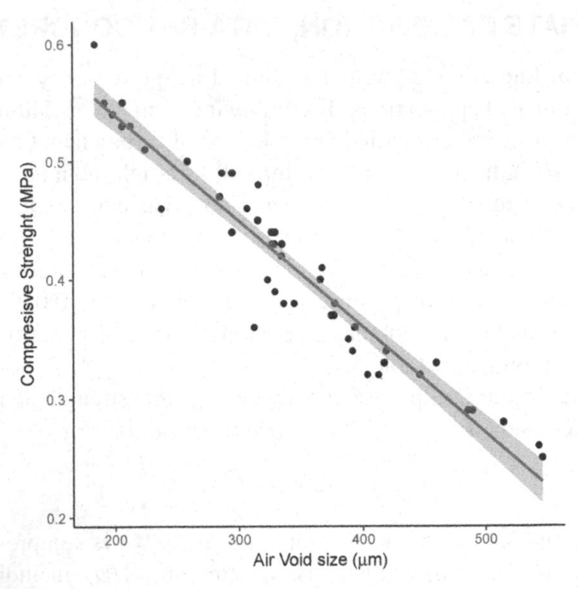

Figure 4.2 Linear regression for air-void size (μm) vs compressive strength (MPa) for the data-set based on the work of Zhang et al. (8). The grayed areas represent a 95% confidence interval in prediction.

compressive strength vs air-void size. Even if both the models present a very good fit, we can see at a glance that *in this specific case*, porosity contribution is bigger in comparison with that of air-void size. A stronger knowledge of the nature of the problems investigated can strengthen this hypothesis. From a statistical point of view, it would be necessary to test the model, as already reported, with an external test set.

4.3.1 BIVARIATE MODELS

Since the CNTs also have an influence on the compressive strength of the concrete, the authors performed a linear regression considering the following formula:

$$y = \beta_0 + \beta_1 \cdot x_1 + \beta_2 \cdot x_2$$

where x_1 represents the porosity (%) and x_2 the CNT content. The regression coefficients, calculated as seen in Chapter 2, are reported for each density in the tables.

Again, as for the univariate linear models, the fitting is good, and looking at the values of the coefficient for x_1 and x_2 we can see that the contribution of CNT *is bigger in comparison with the air-void size*. Also the fit value, *indicates* that porosity and CNTs content have significant influence on low density foamed concrete.

If we would like to visually represent, in that case, we should make use of planes instead of lines, and eventually, as reported in the paper, the model can be further extended, applying a response of surface representations.

Table 4.1
Data-set based on the work of (8).
Carbon nanotubes content (CNTs), compressive strength, density, porosity and air-void of ultra-light foamed concrete are reported.

Density (kg/m^3)	CNT Content *mass%*	Compr. Strength (MPA)cs	Porosity %	Air-void Size (*mu*m)
200	0.00	71.55	546.5	0.25
200	0.05	71.50	514.6	0.28
200	0.10	71.49	489.2	0.29
200	0.15	71.45	445.8	0.32
200	0.20	71.43	417.3	0.33
200	0.25	71.40	394.1	0.36
200	0.30	71.41	391.7	0.34
200	0.35	71.42	403.9	0.32
210	0.00	71.54	543.6	0.26
210	0.05	71.49	485.2	0.29
210	0.10	71.44	459.3	0.33
210	0.15	71.43	418.9	0.34
210	0.20	71.39	312.6	0.36
210	0.25	71.32	329.4	0.39
210	0.30	71.38	376.7	0.37
210	0.35	71.39	388.3	0.35
220	0.00	71.43	412.8	0.32
220	0.05	71.40	393.5	0.36
220	0.10	71.38	377.4	0.38
220	0.15	71.32	328.9	0.43
220	0.20	71.29	316.3	0.48
220	0.25	71.29	315.9	0.45
220	0.30	71.34	334.7	0.42
220	0.35	71.33	336.4	0.38
230	0.00	71.38	374.8	0.37
230	0.05	71.36	365.9	0.40
230	0.10	71.36	367.4	0.41
230	0.15	71.34	335.0	0.43
230	0.20	71.32	326.8	0.44
230	0.25	71.27	287.1	0.49
230	0.30	71.26	285.3	0.47
230	0.35	71.28	294.9	0.44
240	0.00	71.35	344.8	0.38
240	0.05	71.33	327.2	0.43
240	0.10	71.32	329.5	0.44
240	0.15	71.28	294.8	0.49
240	0.20	71.20	206.3	0.55
240	0.25	71.20	205.9	0.53
240	0.30	71.21	224.7	0.51
240	0.35	71.24	238.2	0.46
250	0.00	71.33	323.5	0.40
250	0.05	71.30	307.2	0.46
250	0.10	71.26	259.4	0.50
250	0.15	71.20	198.5	0.54
250	0.20	71.16	183.6	0.60
250	0.25	71.20	191.7	0.55
250	0.30	71.21	212.6	0.53
250	0.35	71.26	295.5	0.49

Table 4.2

Models for compressive strength depending on porosity percentage and compressive strength depending on air-void size.

	Dependent Variable:	
	Porosity	Av
	(1)	(2)
Compressive Strength	−1.101***	−1023.957***
	(0.034)	(46.194)
Constant	71.794***	764.495***
	(0.014)	(19.358)
Observations	48	48
R^2	0.959	0.914
Adjusted R^2	0.958	0.913
Residual Std. Error (df = 46)	0.020	26.764
F Statistic (df = 1; 46)	1065.204***	491.355***
Note:	*p<0.1; **p<0.05; ***p<0.01	

Table 4.3

Models for compressive strength depending on the porosity, air-void size and content of CNT.

	Dependent Variable:	
	cs	
	(1)	(2)
por	−0.665***	−1.023*
	(0.059)	(0.446)
CNT		
av		0.0003
		(0.0004)
Constant	47.840***	73.253*
	(4.204)	(31.682)
Observations	8	8
R^2	0.955	0.960
Adjusted R^2	0.948	0.944
Residual Std. Error	0.008 (df = 6)	0.008 (df = 5)
F Statistic	127.792*** (df = 1; 6)	60.557*** (df = 2; 5)
Note:	*p<0.1; **p<0.05; ***p<0.01	

Table 4.4

Comparison for a model of the compressive strength based on the contribution of porosity + content of CNT and porosity + dimension of air-void size. Models are calculated for a density of 200 kg/m^3.

	Dependent Variable:	
	cs	
	(1)	(2)
por	−0.953***	−1.023*
	(0.084)	(0.446)
CNT	−0.134**	
	(0.036)	
av		0.0003
		(0.0004)
Constant	68.453***	73.253*
	(6.014)	(31.682)
Observations	8	8
R^2	0.988	0.960
Adjusted R^2	0.983	0.944
Residual Std. Error (df = 5)	0.005	0.008
F Statistic (df = 2; 5)	208.306***	60.557***
Note:	*p<0.1; **p<0.05; ***p<0.01	

Table 4.5

Comparison for a model of the compressive strength based on the contribution of porosity + content of CNT and porosity + dimension of air-void size. Models are calculated for a density of 210 kg/m^3.

	Dependent Variable:	
	cs	
	(1)	(2)
por	−0.608***	−0.543***
	(0.083)	(0.115)
CNT	0.003	
	(0.047)	
av		−0.0001
		(0.0001)
Constant	43.785***	39.118***
	(5.970)	(8.168)
Observations	8	8
R^2	0.975	0.977
Adjusted R^2	0.964	0.967
Residual Std. Error (df = 5)	0.008	0.008
F Statistic (df = 2; 5)	95.716***	104.189***
Note:	*p<0.1; **p<0.05; ***p<0.01	

Table 4.6

Comparison for a model of the compressive strength based on the contribution of porosity + content of CNT and porosity + dimension of air-void size. Models are calculated for a density of 220 kg/m^3.

	Dependent Variable:	
	cs	
	(1)	(2)
por	−1.190***	−1.308
	(0.190)	(1.003)
CNT	−0.135	
	(0.079)	
av		0.0005
		(0.001)
Constant	85.338***	93.582
	(13.541)	(71.089)
Observations	8	8
R^2	0.921	0.879
Adjusted R^2	0.890	0.830
Residual Std. Error (df = 5)	0.017	0.021
F Statistic (df = 2; 5)	29.278***	18.133***
Note:	*p<0.1; **p<0.05; ***p<0.01	

Table 4.7

Comparison for a model of the compressive strength based on the contribution of porosity + content of CNT and porosity + dimension of air-void size. Models are calculated for a density of 230 kg/m^3.

	Dependent Variable:	
	cs	
	(1)	(2)
por	−1.043*	−0.404
	(0.408)	(0.993)
CNT	−0.111	
	(0.154)	
av		−0.0005
		(0.001)
Constant	74.868*	29.377
	(29.149)	(70.414)
Observations	8	8
R^2	0.864	0.854
Adjusted R^2	0.810	0.796
Residual Std. Error (df = 5)	0.017	0.017
F Statistic (df = 2; 5)	15.943***	14.668***
Note:	*p<0.1; **p<0.05; ***p<0.01	

Table 4.8

Comparison for a model of the compressive strength based on the contribution of porosity + content of CNT and porosity + dimension of air-void size. Models are calculated for a density of 240 kg/m^3.

	Dependent Variable:	
	cs	
	(1)	(2)
por	−1.206***	−2.618*
	(0.239)	(1.074)
CNT	−0.213	
	(0.121)	
av		0.002
		(0.001)
Constant	86.471***	186.545*
	(17.049)	(76.205)
Observations	8	8
R^2	0.908	0.904
Adjusted R^2	0.871	0.865
Residual Std. Error (df = 5)	0.020	0.021
F Statistic (df = 2; 5)	24.639***	23.489***
Note:	*p<0.1; **p<0.05; ***p<0.01	

Table 4.9

Comparison for a model of the compressive strength based on the contribution of porosity + content of CNT and porosity + dimension of air-void size. Models are calculated for a density of 250 kg/m^3.

	Dependent Variable:	
	cs	
	(1)	(2)
por	−1.060***	−1.308***
	(0.105)	(0.296)
CNT	−0.009	
	(0.049)	
av		0.0003
		(0.0003)
Constant	75.994***	93.603***
	(7.497)	(21.017)
Observations	8	8
R^2	0.968	0.973
Adjusted R^2	0.956	0.962
Residual Std. Error (df = 5)	0.013	0.012
F Statistic (df = 2; 5)	76.175***	88.580***
Note:	*p<0.1; **p<0.05; ***p<0.01	

4.4 MULTIVARIATE CALIBRATION

In the previous section, we've seen that starting from a univariate and bivariate regression, we can study our system using an increasing number of variables transitioning in the field of multivariate regression. In order to better explain how to deal with this subject, we will get back to spectroscopy as in the beginning of the chapter.

We've seen that the first natural approach while dealing with spectroscopy data in order to quantify a response is to make use of the Lambert-Beer Law. However, what can we do when the peaks in the spectrum overlap, and it is not possible to resolve the peaks? Due to this occurrence, it is not usually possible to use absorbance at a single wavelength, and so we need to extend the range of wavelength employed considering *portions* or the whole spectra.

In order to extract as much information as possible from our data-set and knowing that spectral data are highly correlated (i.e., that we have high correlation among wavenumbers of the spectra), we will again perform a *regression*, but this time instead of an x, we will have a matrix X, and analogously the y will be a vector Y. We will try to describe in detail the most common methodologies employed, but before that, we will describe the general idea behind two of the most widely applied: the principal component regression and partial least square.

In both techniques we try to find a few *linear combinations*, *components*, or *factors* of the original x values and use them in the regression equations. In matricial form we have:

$$X = TP^t + E \tag{4.1}$$

$$y = Tq + f \tag{4.2}$$

where the matrix P and the vector q are called *loadings*, and analogously to the PCA, describe how the original variables in T relate to the original matrix X and vector y. E and f represent the noise of X and y, respectively. PCR and PLS use *different criteria* for the computation of the matrix T, while the *rest of computation* is *identical.*

4.4.1 PRINCIPAL COMPONENT REGRESSION

One of the methods to *compress* the information included in the data-set and to reduce the correlation among variables (*multicollinearity*) is to calculate the principal component of the data-set before applying the regression. Due to the definition already presented of *principal components*, we can also write that we try to find few linear combinations of the x-values, and use them in the regression process. The general equations are the one already reported in 4.1 and 4.2.

In PCR, the estimated scores matrix T consists of the most *dominating* component of X, and the corresponding matrix of loadings is denoted by P. As seen in the previous chapter, the principal components *can be computed using the eigenvector obtained via SVD.*

We need to clarify why in the previous definition we wrote *dominating* components. We can use the same definition used in the PCA and so the components that explain the *maximum variance* but also using *t-tests*.

As a note, if the number of samples is larger than the number of variables, the maximum number of components that we can extract is the number of variables. Keep in mind that in material science and especially while using spectroscopic techniques, this means having thousands of samples, and this is rarely the case in preliminary studies that does not involve combinatorial approaches. Finally, PCR is *not invariant* to changes in the scale of X, and neither is PLS.

Even if we can use the raw acquired data for PCR and PLS (and sometimes it can be required, where we do know that selected variables contribute to the need to "weigh" more in the model), generally the common practices employed in a multivariate workflow require a preprocessing step that are the same as those reported in the chapter dedicated to clustering analysis.

4.4.2 AN EXAMPLE OF MULTIVARIATE REGRESSION USING THE GASOLINE DATA SET

The following analysis is based on the tutorial article presented by Mevik and Wehrens (5) about the package for **R**, **pls** We will refer to their analysis of the dataset gasoline (2). The data-set is composed of 60 NIR spectra measured at 401 nm wavelength at 2 nm interval (900-1700 nm) of gasoline sample (X) and their number of gasoline (Y). We will start calculating the PCR which is straightforward thanks to the package (see appendix for the calculations). In detail, the characteristic of the models are the following:

In Figure 4.3, we reported the root mean squared error of prediction (RMSEP) using both the cross-validated estimate and the adjCV bias-corrected CV estimate (4) (that, as expected, shows no difference for a leave one out cross-validation).

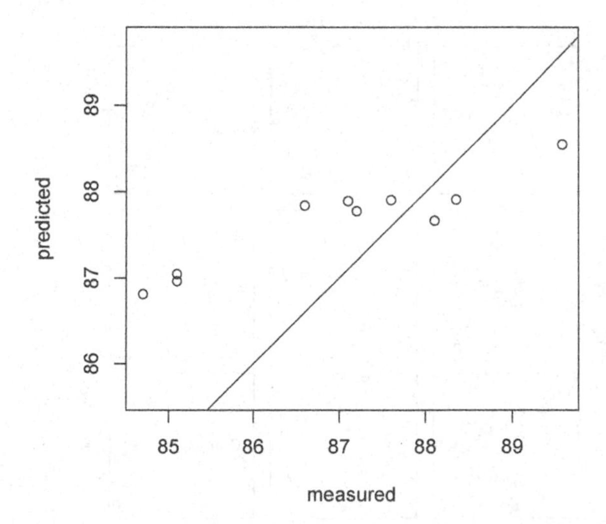

Figure 4.3 Cross-validated RMSEP curves for the gasoline data. Both ordinary cross-validation (continous line) and a bias-corrected CV (dashed) are reported.

Table 4.10
Explained variance depending on the number of components for PCR model calculated on the gasoline data-set.

Comp 1	Comp 2	Comp 3	Comp 4	Comp 5	Comp 6	Comp 7
79.859	8.264	5.417	3.003	1.196	0.640	0.369

From the results reported both in 4.10 and 4.3, it is possible to see that even considering just two components is sufficient for explaining almost 86% and 97%, respectively, of the X and Y matrix previously defined. So, selecting three components should be a safe choice for our data-set.

In Figure 4.4, we reported the scatter plots referred to the *first three scores* for the model calculated. Analogously to what we've seen in the PCA chapter, PCA scores are used to search for patterns among the samples. At a quick glance, we can define two clusters, thanks to PC1 vs PC3, but there is no clear pattern in the data.

The loading plot (4.5) highlights the contributions for the first and second components. As reported in the figure, the first loading explains 79.9% of the variance of the data-set, while the second component accounts for 8.3%. Loadings plot, again, analogously at what we've seen in the PCA session can be used to search for known peaks value by the user in order to relate which frequencies "weigh more" in calculating the model. Considering the first loading, the frequencies that weighed more in order of magnitude are in the interval between 1600-1700 nm. We can observe a peak near 1150 nm, another peak near 1200 nm, and finally at 1400 nm.

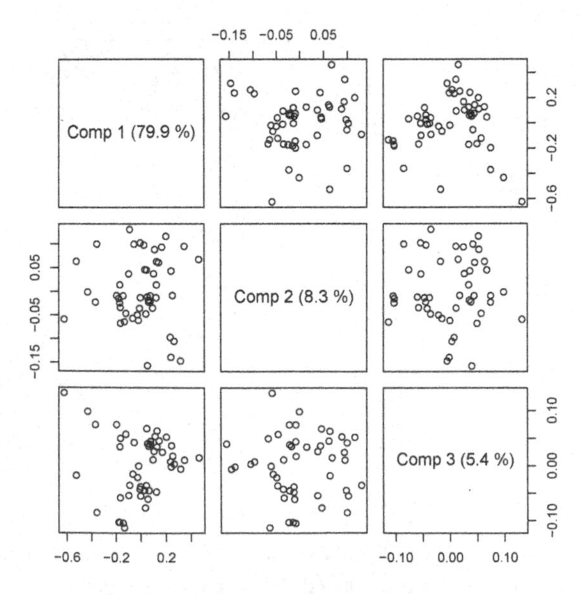

Figure 4.4 PCR score plot for the data-set gasoline (2).

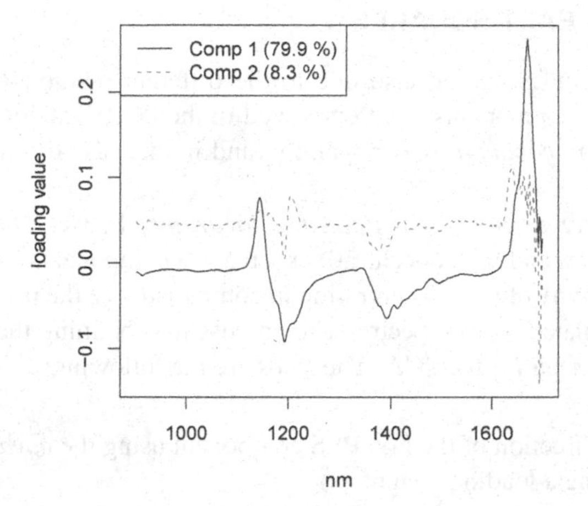

Figure 4.5 Black continuous line reports the loading value for the first component, while the dashed lines represent the loading for the second component.

Table 4.11

	octane.(Intercept)	comp 1	comp 2	comp 3	comp 4
test	1.537	1.323	1.257	0.463	

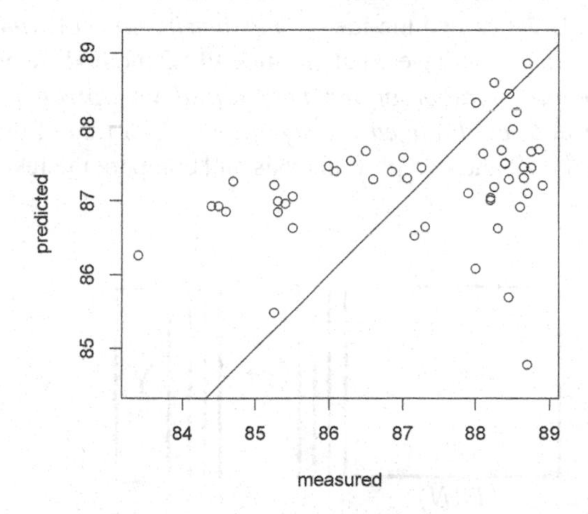

Figure 4.6 Cross-validated prediction of the gasoline data-set obtained by PCR modelling.

Finally, in Figure 4.6 and Table 4.11, we present the results of the model in predicting the response of a test set. Two components give us an error value in prediction of *1.323*.

4.4.3 PARTIAL LEAST SQUARES

The development of PLS started also due to the different methodologies that we can use to select the components to be employed in the PCR. The idea was that we can employ factors *in common* by our x and y (and in this case the spectra and the concentration).

In other words, we can try to *maximize the covariance* between the linear combination of the spectra and the concentrations, creating components which are *more related to the variability* of the concentration in comparison, as the principal components are only calculated on the spectra. The process for obtaining the PLS components is iterative, as seen for the PCA. The steps are the following:

- calculate the direction of the first PLS component using the *covariance criterion* obtaining a loading weight
- compute the scores along the axis found
- perform regression of the variables X on the computed scores obtaining the loading vector
- perform the regression of y on the scores along the axis
- calculate the residuals, subtract them from the original matrix X and vector y
- keep iterating till a stop criteria is reached

The note reported previously for the PCR that the number of samples vs number of variables is still valid for PLS, and so if the number of samples is larger than the number of variables, the maximum number of components that we can extract is the number of variables. If we extend the idea of *maximizing the covariance to several y-variables simultaneously*, we have what is called PLS2 method. Keep in mind that while *increasing the interpretation of our data*, it *trades off prediction power*, and so it can be still useful *to calibrate each y separately*. As a rule of thumb, when in doubt, perform both a pls2 and two pls1 analysis and compare the results.

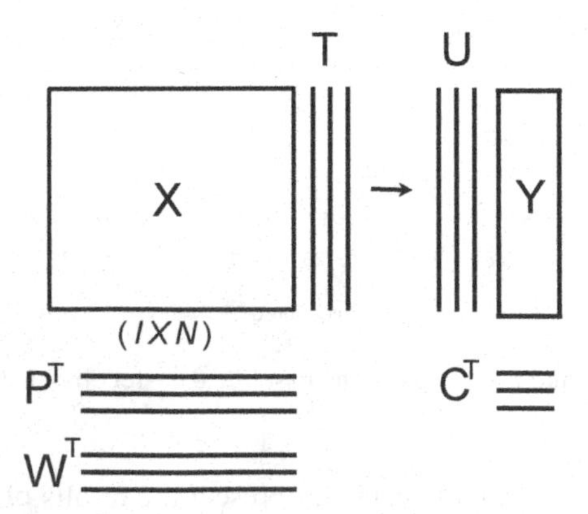

Figure 4.7 Schematic diagram for PLS. Data can be arranged in two tables, X and Y, based on the work of (7).

As reported by (7), PLS regression algorithms tolerate moderate amounts of missing data.

There are algorithms that work with the original data matrices X and Y or with the preprocessed original matrix (scaled and centered) but also there exist *kernel algorithms* that can work with variance-covariance matrices $X'X$,$Y'Y$ and $X'Y$ or *association matrices* XX' and YY'. For all the algorithms considered, we have in common the *orthogonality* between the model component, good summarizing properties of the X-scores, and interpretability of the model parameters. Analogously as what we've seen for the principal component analysis also PLS can be solved with the aid of NIPALS algorithm. The steps followed are:

- get a starting vector of u (usually a column of Y)
- calculate the X-weights w $X'u/u'u$ and then normalize w to 1.0
- calculate x-scores according to $t = Xw$
- calculate the weights $c = Y't/t't$
- update the set of Y-scores $u : Yc/c'c$
- test convergence on the scores, checking if the difference between the scores differers from a previous iteration of the value less than a predefined tolerance ε, and if convergence is not reached, recalculate the weights
- if convergence was reached in the previous step, remove the component from the starting matrix X and Y and start from the begining with the newly obtained matrices
- continue extracting the component until cross validation indicates that there is no more significant information in X about Y

Now we apply the partial least square methodology (again without preprocessing) to the gasoline data-set, obtaining the following results. RMSEP plot shows that the two components can explain almost 86% of the X variance and 97% of the Y variance.

In Table 4.8 and Figure 4.9, we reported the score plot and loading plot, respectively.

Score plot does not highlight any clusters. Loadings plot help us to highlight the frequencies that carry more information. In order of magnitude, for this data-set we have the interval between 1600-1700 nm, a peak near 1150 nm, another peak near 1200 nm, and finally 1400 nm, as previously seen for the PCR model. The plot of Table 4.10 highlights the lowest error in prediction in comparison with the PCR model previously calculated.

Can we try to obtain a better model?

Since we know that spectroscopic data can contain a source of noise (artifacts) related to the scattering effect, we can rely on an effective technique for preprocessing data, called Multiplicative Scatter Correction filter. You can find the details to this methodology and other preprocessing methods in the work by Mevik (5).

The results obtained applying the preprocessed data are the following:

Analogously to what we've seen in the previous PLS model, a few components let us explain the variance of X and Y. We have a slight improvement with almost 93% of X variance and 98% of Y variance explained with two components. The loadings

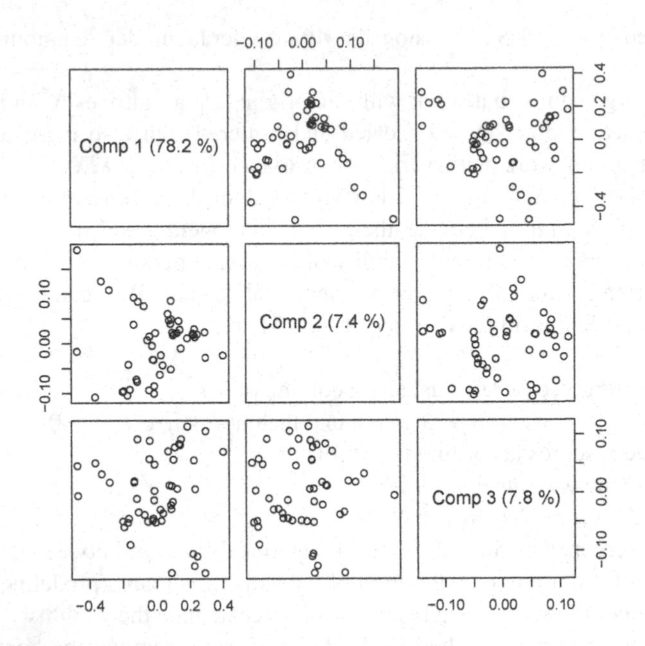

Figure 4.8 PLS score plot for the data-set gasoline (2) obtained without preprocessing the data.

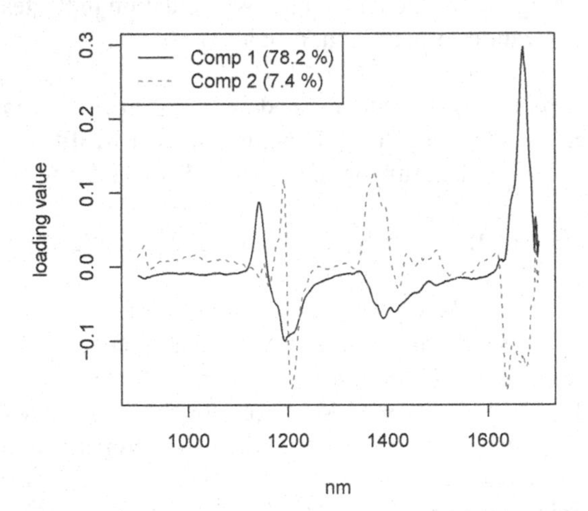

Figure 4.9 Black continuous line reports the loading value for the first component, while the dashed lines represent the loading for the second component. No preprocessing was performed.

plot highlights almost the same intervals, but with an increased resolution (sharper peaks) in the interested zones, and finally the prediction plot shows a good prediction power on the chosen train set.

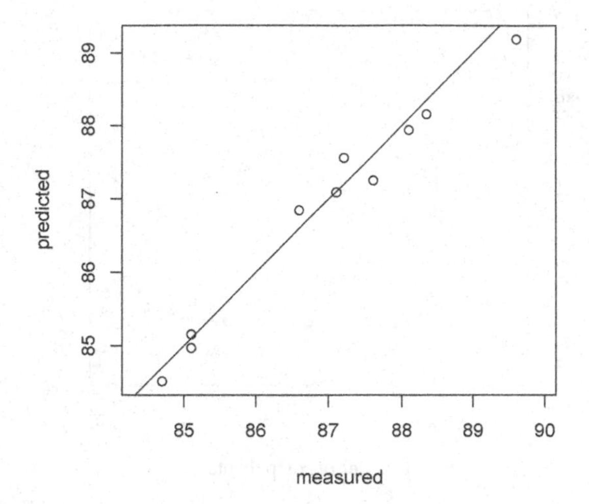

Figure 4.10 Measured vs predicted value for the model obtained via PLS regression without preprocessing data on the data-set GASOLINE.

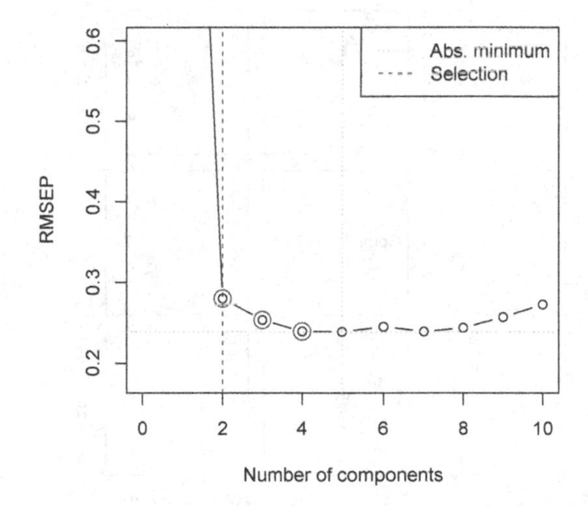

Figure 4.11 Number of components vs RMSEP for the model obtained after preprocessing using MSC on the data-set GASOLINE.

As a final note, comparing the data obtained without preprocessing, we've confirmed that the *PLS* (this is not *always* the case) uses less components to express the same variance in comparison with *PCR*. Score plots of both models did not highlight any clustering. The loading plots highlight the most important variables (wavelengths) that had the highest contribution in the calculations of the models, and highlighted the same intervals of frequencies, and both models perform well in prediction, as seen by the value of RMSEP.

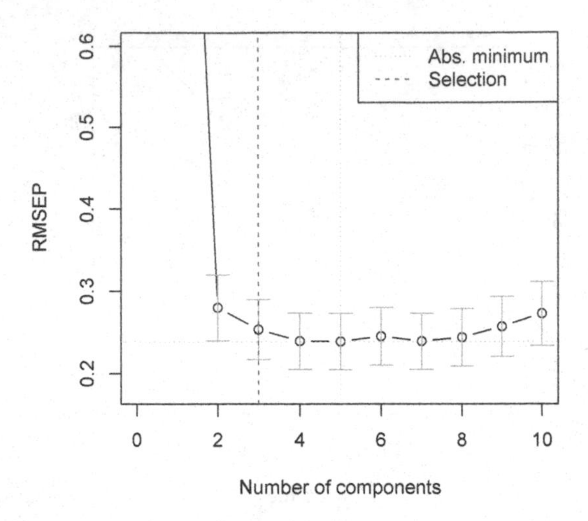

Figure 4.12 Number of components selected using permutation methodology.

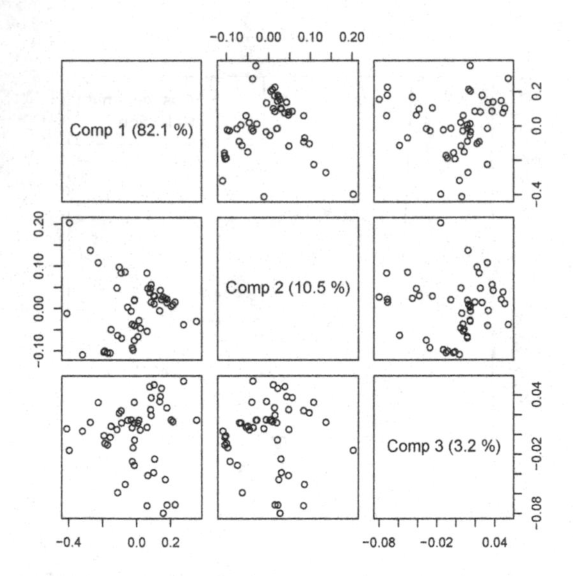

Figure 4.13 Score plot of first three components for the model obtained after preprocessing using MSC on the data-set GASOLINE after cross-validation on training set.

4.5 OTHER REGRESSION METHODOLOGIES

4.5.1 NWAY METHODOLOGIES

Thanks to the progress of analytical techniques and the dedicated acquisition hardware, the data-set that we gather are starting more and more to include a third dimension related to the trends in time of the monitored properties.

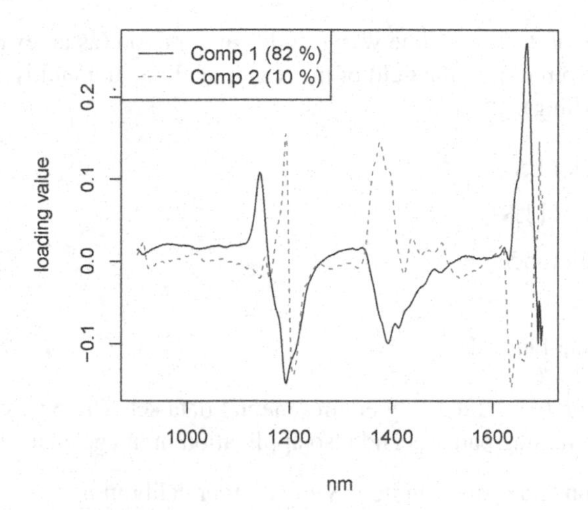

Figure 4.14 Loading plot of first three component for the model obtained after preprocessing using MSC on the dataset GASOLINE after cross validation on training set.

As an example we can think of coupled techniques calorimetric and spectroscopy techniques as TGA-FTIR where we acquire Infrared Spectra while performing a Thermogravimetric Analysis, Nuclear Magnetic Resonance data NMR data acquired in multidimension and at different time, Cromatography and Mass Spectroscopy, etc.

Calibration in these cases will take care of trying to predict a parameter for an unknown sample after relying on a data-set trained in more than two dimensions.

It is due to notice that among the multiway data-set, as already mentioned in principal component and cluster analysis, we can include all the data-sets obtained from techniques that involve acquiring images in time.

Several techniques are available in order to deal with this specific kind of data. What do they have in common?

- they involve an *unfolding* step of our data
- several nway methodologies are an extension in nway of widely used calibration methods
- they rely on *loadings* and *score*

What are the applications?

Depending on the arrangement of the block of information gathered and also if we are going to use the block of information for prediction or optimization problems, we can rely on different methodologies.

- Two way multiblock data
- Exploring and predicting

- Three blocks of data, with one way (mode) in common (as an example we have data gathered from the field of process modelling and multivariate statistical modelling):
    - Multiblock PLS
    - Hierarchical PLS
    - Latent path modelling
- Three-way one-block data
    - Exploration (e.g., three way environmental data-set. Three way analysis helpful in finding patterns and also application in image analysis).
    - Calibration and resolution item Second order calibration
- Three-way two block data
    - example Modelling and multivariate statistical process
    - example Multivariate calibration and control chart
- Three way multiblock
    - example multivariate statistical process control of batch processes
- Four data and more Image analysis
  In the suggested literature section, readers can find more references about multiway analysis.

4.5.2 A SHORT HISTORY OF PARTIAL LEAST SQUARES

Partial Least Squares started from the computational ideas of Herman Wold in the 60s, developed with his research group at the Uppsala University (Sweden). Wold worked on different procedures starting from a power method for computing Principal Components, that gave origin to *"Non Linear Iterative Least Squares"* and to the *Nipals Procedures*.
In the early 70s, these procedures were developed in order to include path models with latent variables (models associating two blocks of variables) and changed their names from *Nipals Procedures* to *Nipals Modelling*. In the mid 70s, a new methodology was developed in order to deal with three latent variables and handling more than one between-block relation, while in the end of the 70s, NIPALS is shortened to PLS and other types of econometric models are included in the developing of PLS.
During the 80s, Svante Wold (son of Hermald Wold), under his father's suggestions, started working on the application of PLS framework, for regression analysis with chemical data. Svante Wold and Harald Martens and Herman Wold published the first papers on PLS and multivariate calibration in 1983, and the analytical chemistry community was informed soon after. Improved and modified during the middle 1980s, the PLS start getting adopted due to a need for analytical tools capable of dealing with multicollinearity, missing values, and large data sets.
Svante Wold also became a pioneer and leading figure in the newly named field of chemometrics, focusing on more industrial aspects and the computational aspect.
In the 90s, Svante Wold was promoting PLS within Chemical industries and also Tormod Naes and Harold Martens further develop the methods and expanded it to new directions. PLS software started to be more widely adopted thanks to Umetrics and Unscrambler. Also, SAS added it to the PLS-R procedure, attracting more researchers in the field.
For a detailed story of the methodology, the readers can consult the exhaustive reference written by Gaston Sanchez at the website https://sagaofpls.github.io/, author of the book "PLS Path modelling with R" (6).

4.5.3 Q.A.

Q. Should I use PCR or PLS?

A. This is probably the most asked question while presenting PCR and PLS. Again, as you have imagined, there is not an unique answer, and it also depends on who you ask. Some scientists are not interested in the answer, while others have embraced religious wars on the topic. If we leave out the computational cost, which does not matter anymore due to the continual progress of hardware, it varies depending on the following points:

- you are using the technique in order to check if a sample belongs to a specific class of a data-set using a multivariate approach
- always try to use an external test set to check for the validity of a model

Q. What is the difference between MLR and PLS?

A. PLS assumes the presence of a *few underlying variables* in our system that influence it. Also, they are related to both X and Y, and we assume that the X and Y are not independent of each other. If the number of *latent variable* is equal to the number of the *X-variables* in the system, PLS and MLR give identical results.

Q. What preprocessing methodology should I use?

A. The advice is the same given for PCA in Chapter 3. The idea is still to maximize the info without increasing the noise in the data-set. Centering, autoscaling, SNV, Multiplicative Scattering Correction, Selection of Intervals, etc., are all tools that can be used to enhance the quality of our modelling.

Q. How many samples?

A. Again it depends on the variables you are investigating. It is not unusual to see spectroscopic data-set (so thousands of variables) while dealing with at least thirty samples even if this is, of course, not optimal. Hundreds would be more suitable to make the data-set more robust and the more the better.

4.5.4 ESSENTIAL REFERENCES

One of the references that I always use is *Multivariate Calibration and Classification* by Naes et al. because of its rigorous, concise, and clear approach.

As previously cited, I would suggest the reader to have a look at the calibration section of the book *"Practical guide to Chemometrics"*, by Brereton, which is detailed and clear (for a chemist).

A seminal article about PLS by Svante Wold et al. is "PLS-Regression, a basic tool of chemometrics", published in Chemometrics and Laboratory system (7).

For a reference about how to perform PLS calculations using the R language, readers can rely on the package from (5) and also (1).

As already reported for principal component analysis, researchers and practitioners in the field of Material Science will probably have to deal with complex numbers. The article by D. Rodrigues et al. "Complex numbers-partial leastsquares applied to the treatment of electrochemical impedance spectroscopy data" will be helpful in presenting solutions to these problems.

Bibliography

1. Heide Garcia and Peter Filzmoser. Multivariate Statistical Analysis using the R package chemometrics. *Vienna: Austria*, pages 1–71, 2011.
2. John H. Kalivas. Two data sets of near infrared spectra. *Chemometrics and Intelligent Laboratory Systems*, 37(2):255–259, jun 1997.
3. Harald. Martens and Tormod. Næs. *Multivariate calibration*. Wiley, 1989.
4. Bjoørn Helge Mevik and Henrik René Cederkvist. Mean squared error of prediction (MSEP) estimates for principal component regression (PCR) and partial least squares regression (PLSR). *Journal of Chemometrics*, 18(9):422–429, sep 2004.
5. Bjørn Helge Mevik and Ron Wehrens. The pls package: Principal component and partial least squares regression in R. *Journal of Statistical Software*, 18(2):1–23, 2007.
6. Gaston Sanchez. *PLS Path Modeling with R*. Trowchez Editions, 2013.
7. Svante Wold, Michael Sjöström, and Lennart Eriksson. PLS-regression: A basic tool of chemometrics. In *Chemometrics and Intelligent Laboratory Systems*, 2001.
8. Jing Zhang and Xiangdong Liu. Research on the influence of carbon nanotubes (CNTs) on compressive strength and air-void structure of ultra-light foamed concrete. *Mechanics of Advanced Materials and Structures*, pages 1–8, dec 2018.

5 Case Studies

In the following section we will present three full papers that report detailed case studies in the field of multivariate techniques applied to material sciences.

The first one is a contribution on the application of factorial *design applied to the fabrication of a micro/nano hierarchical ZnO superhydrophobic surface via simple and fast chemical bath deposition method without hydrophobization post-treatment*. Thanks to the design applied, it was possible to assess the most effective parameters (additive type, seed layer solution concentration, and initial concentration of Zn (II) ions), and find an optimum level of parameters used to obtain good superhydrophobic characteristics of ZnO surfaces. The difference between experimental and estimated results was also evaluated, and confirmed the predictability, reproducibility, and high accuracy of experimental data (with an error of 6.5%).

The second contributed paper presents the *application of evolutionary design and Successive Bayesian Estimation in order to introduce the reader to advanced DOE methodologies* in comparison with the one presented in Chapter 2. It demonstrated how joint application of EDOE and SBE methods provides evident advantages for prediction of the rubber lifespan developing a system of models coherently describing various rubber properties. In the article the authors consider that the existence of such a system, which correctly describes experimental data, proves possibility and validity of the extrapolation.

The third reports an interesting *application of how a very simple technique, such as Principal Component Analysis, was applied to analyze the data obtained through sweep cyclic voltammetry, allowing the study of the influence of selected chemical inhibitors on the morphology of the voltammograms*. PCA was able to highlight clusters depending on the pH and the inhibitor admixture used, and also to provide more information in specific areas of the voltammogram that were not visible via a "classic" analysis workflow.

5.1 FAST FABRICATION OF ZNO SUPERHYDROPHOBIC SURFACES WITHOUT CHEMICAL POST-TREATMENT: INVESTIGATION OF IMPORTANT PARAMETERS USING TAGUCHI MIXED LEVEL DESIGN L8 (4^1 2^3)

R. Norouzbeigi[1,*] and E. Velayi[1]

5.2 INTRODUCTION

Construction of superhydrophobic coatings on various materials, such as metallic foils, textiles, and mesh surfaces has become one of the most interesting subjects in recent years (23; 49; 30). Superhydrophobic surfaces have numerous industrial applications in various areas, such as self-cleaning (29), anti-icing (45), anti-corrosion (53; 10; 51), anti-fogging (45), and oil/water separation (5; 22; 28).

Surface chemistry and topographical structure are two main factors determining the wettability of the solid surface (55).

Literature surveying reveals that superhydrophobic surfaces can be fabricated by modifying the rough surface with low surface energy materials or creating a suitable roughness on the hydrophobic surface. A variety of methods have been used for fabrication of superhydrophobic surfaces, including sol-gel, chemical vapor deposition (27), hydrothermal (47), layer-by-layer technique, and chemical bath deposition (CBD) (38), resulting in water contact angle (WCA) greater than 150, and contact angle hysteresis (CAH) lower than 10 (30; 32; 38; 49). In this way, the solution immersion method was employed to produce a superhydrophobic gibbsite film on aluminum foils (38). This method includes deposition of gibbsite film with a binary micro/nano hierarchical structure on the substrate, and then modification of the surface with stearic and palmitic acids as surface free energy reducing materials. Wang et al. (51) (19) prepared superhydrophobic Pb-coated titanium alloy surfaces via the solution immersion method, followed by a post-hydrophobization using fluoroalkylsilane (FAS). In another study, Song and coworkers (44) fabricated three-dimensional (3D) flower-like Fe_3O_4 micro/nanoflakes on iron plates by the hydrothermal synthesis method. Vinyl triethoxysilane was applied as a low surface energy material to obtain the superhydrophobicity. Up till now, low surface materials have been employed routinely to yield superhydrophobic characteristics of metal oxide surfaces. Toxicity, high dissolution rate of modifiers in common solvents, and their poor resistance in the harsh conditions are the most challenging problems for practical application of superhydrophobic surfaces (27; 11; 14; 21).

Recently, some scientists have reported the fabrication of superhydrophobic surfaces without future modification step (27; 47; 11). However, most of them have extensively used one factor at the time approach for investigating the effect of process

[1]Nanomaterials and Surface Technology Research Laboratory, School of Chemical, Petroleum and Gas Engineering, Iran University of Science and Technology, P.B. 16765–163, Narmak, Tehran, Iran.
*Corresponding author: norouzbeigi@iust.ac.ir

parameters on the surface wettability. These parameters are including the precursor concentration, time and temperature of reaction, molar ratio of reactants, type and concentration of modifier, and so on. Evaluation of these parameters separately is too time-consuming and expensive. Moreover, a great number of experiments are required to determine the optimal conditions. Statistical tools such as the design of experiments could be used to evaluate a different number of independent parameters on the proposed response without doing a large number of experiments (33; 40). Therefore, reduction of the experimental runs, decreasing the overall costs, and determining the effective parameters during a short time are important advantages related to the design of experiments. Several techniques can be applied to optimize the fabrication procedure, such as Taguchi, Box-Behnken, central composite design, and so on (36). Taguchi design is a simple, economical, and efficient statistical tool widely employed for optimization of multi-factor processes in a large number of diverse fields, such as electronics (17), material science (6; 41; 20; 48), medicine (7), and mechanical engineering (2; 35; 36). In this method, mathematical and statistical techniques are combined and utilize orthogonal arrays in order to investigate a large variety of variables with a small number of experiments (40; 46). Orthogonal array presents the subset of the full factorial design. The interactions between parameters are neglected and the impact of each variable can be evaluated regardless of other factor effects. In order to analyze the results, factor main effect and statistical analysis of variance (ANOVA) can be applied to evaluate the impact of selected influential parameters and estimate the optimal levels of controlling parameters. The Taguchi approach involves the following steps- 1) choice of controlling parameters influencing the experimental response and selection of their levels; 2) design of experiments to study the impact of the parameters on the process; 3) running of the experiments; 4) analysis of the process response according to the selected output; 5) running of the confirmation test at optimal conditions to determine the accuracy and reproducibility of the experiments. In this paper chemical bath deposition method was used to fabricate the superhydrophobic ZnO surfaces by the one-stage procedure without post-modification by low surface energy materials, and the fabrication parameters including initial concentration of Zn(II) ions, type of additive, seed layer solution concentration, and drying temperature were studied and optimized in terms of wettability properties of zinc oxide surfaces using Minitab 16 software with Taguchi mixed design L8 (4^1 2^3).

5.3 MATERIALS AND METHODS

5.3.1 MATERIALS

Pure stainless steel 304 L mesh with the mesh number of 50 and dimensions of 25 mm × 25 mm were used as substrates. Zinc(II) nitrate hexahydrate ($Zn(NO_3)_2 6H_2O$), Hexamethylenetetramine ($C_6H_{12}N_4$), Urea (CH_4N_2O), Ethanol (C_2H_6O), Acetone (C_3H_6O), and polyvinylpyrrolidone (PVP) were supplied from Merck Company. Polyethylene glycol 6000 (PEG 6000) was purchased from Biotech Company. All reagents were used without further purification.

5.3.2 DESIGN OF EXPERIMENTS (DOE)

The effect of deposition parameters was studied considering *initial concentration of Zn(II) ions, type of additive, seed layer solution concentration, and drying temperature* on the wettability behavior of zinc oxide surfaces. The seed layer solution concentration was selected at four levels and the other factors were two-leveled. The WCA value was considered as a response variable of the experiments. Table 5.1 shows the selected factors and the levels. These levels were chosen according to the literature review and preliminary experiments. Taguchi designed experimental runs in 5.2 which shows the combination of factors and levels.

Table 5.1
Factors and levels considered for Taguchi mixed design L8 ($4^1\ 2^3$).

Factor		Allocated Letter	Levels
Seed layer solution concentration (mM)	A	A1	0
		A2	20
		A3	50
		A4	100
Initial concentration of Zn(II) ion (M)	B	B1	0.1
		B2	0.2
Drying temperature (°C)	C	C1	100
		C2	150
Type of additive	D	D1	CTAB
		D2	PVP

Table 5.2
Taguchi L8 ($4^1\ 2^3$) array design matrix.

Sample	Factors and Levels Combination			
	A	B	C	D
1	A1	B1	C1	D1
2	A1	B2	C2	D2
3	A2	B1	C1	D2
4	A2	B2	C2	D1
5	A3	B1	C2	D1
6	A3	B2	C1	D2
7	A4	B1	C2	D2
8	A4	B2	C1	D1

5.4 SAMPLE PREPARATION

The stainless steel mesh (substrate) in the dimension of 3 cm × 3 cm was cleaned ultrasonically for 10 minutes with deionized water, ethanol, and acetone, respectively. Then the substrates were chemically etched by nitric acid solution (4 M) for 15 minutes and dried in the ambient temperature. The etched substrate was then immersed in a glass beaker including an aqueous solution of zinc nitrate with different concentrations (0.1 M and 0.2 M), HMTA (HMTA/Zn molar ratio 1:1), and the organic additive with a concentration of 0.1 mM (PEG and PVP). The beaker sealed and placed in the oil bath remained at a temperature of 90±5°C. The synthesis reaction was considered 1 hour for all experimental runs. After the deposition step, the substrate was removed from the solution, rinsed with distilled water and ethanol, and then dried in electrical oven at different temperatures (100°C or 150°C) for 4 hours.

5.4.1 CHARACTERIZATION

The wettability properties of the synthesized ZnO surfaces were determined by measuring of WCA and CAH according to the sessile drop method using a digital optical microscope (DINOLITE, model AM-4113ZT, Taiwan). The contact angles measured during expanding and retracting volume of droplets were considered as advancing and receding contact angles using droplet volume (5μl) oscillation method. CAH (∘) was calculated by subtracting the value of receding angle from the advancing one. The WCA and CAH were measured at five different points for each sample. The mean values and standard deviations are reported for all samples. It should be noted that, the average WCA values of both left and right sides of the droplets were considered for each measurement due to the existence of the micro/nano defects on the superhydrophobic surfaces (37). The microstructure and morphology of ZnO surfaces were characterized by scanning electron microscopy (SEM, Tuscan, model VEGA2) operating at 30.0 kV. The X-ray diffraction analysis of the optimum sample was carried out using X-ray diffractometer (X, Pert pro, model Panalytical, Netherlands) with nickel-filtered Cu Kα radiation (λ= 1.5418). The chemical feature of the sample was evaluated by Fourier transform infrared spectrometer (ATR-FTIR, model ATR-8000, Shimadzu, Japan).

5.5 RESULTS AND DISCUSSION

5.5.1 DOE ANALYSIS

Table 5.3 represents the measured water contact angles of all samples and the contact angle hysteresis for superhydrophobic surfaces. As it is clear from the table, all samples exhibited hydrophobic characteristics with WCA greater than 90°, except samples 2 and 6. The superhydrophobic behaviors were observed only in the case of samples 4 and 5 (WCA>150°, CAH<10∘). The samples which were prepared in the solutions including CTAB showed higher WCA than the others. The standard deviation of measurements is another major output that can give important information about the uniformity of the coating (32). This parameter plays a crucial role in the practical applications, especially for oil/water separation (56). Data in Table 5.3

Table 5.3
Taguchi L8 array design matrix and obtained results.

Sample	Factors and Levels Combination				Response			
	A	B	C	D	Yi	WCA	STD	CAH
1	A1	B1	C1	D1	Y1	145.3±7.0	5.7	
2	A1	B2	C2	D2	Y1	96.5±7.8	6.2	
3	A2	B1	C1	D2	Y1	45.8±4.7	3.7	
4	A2	B2	C2	D1	Y1	150.3±1.9	1.5	4.5±1.1
5	A3	B1	C2	D1	Y1	150.2±5.4	4.3	5.5±0.8
6	A3	B2	C1	D2	Y1	99.3±1.0	0.8	
7	A4	B1	C2	D2	Y1	37.7±4.53	2.3	
8	A4	B2	C1	D1	Y1	137.6±6.3	3.0	

confirms the uniformity of synthesized ZnO coatings. Reported STDs are suitably low in the range of 0.8 and 6.2 .

The results for the main effect of all factors and their analysis of variance anova (ANOVA) are summarized in 5.4 and 5.5. It can be observed that type of the additive (factor D) has a noticeable effect on the hydrophobicity of the surface. Factor A (seed layer solution concentration) is the second important factor. It implies that the organic additives can be attached on a certain plane of ZnO and promote the preferential orientation growth. It leads to the hierarchical structures formation. The topographical structure of surfaces has an important role on their wettability properties. So it is expected that the organic additives influence the hydrophobicity (42; 8; 50). After factor A, the concentration of seed layer (B) plays an essential role in obtaining superhydrophobic ZnO surfaces. An increase in the seed layer solution concentration can increase the seed layer thickness accordingly, and the nucleation sites on the seed layer surfaces can be reduced (18; 25). These phenomena can influence the density of ZnO rods and the surface average roughness. The initial concentration of Zn(II) ions has little effect on the hydrophobic properties compared to the others and finally the drying temperature may be considered as a negligible one. The main effect plot is shown in 5.2. The figure displays the optimum combination of the factor levels in order to obtain higher water contact angles. This combination contains A3 (seed layer concentration of 50 mM), B1 (initial concentration of Zn(II) ions of 0.2 M), C2 (drying temperature of 140°C), and D1 (the additive type of CTAB). It's due to notice that statistically, factor C (drying temperature) is a non-significant parameter (see 5.5).

As the combination of optimal levels of factors was not similar to any of the experiments in Taguchi mixed design (L8 (4^1 2^3)), the extra sample was prepared (sample 9). The estimated WCA was calculated using the following equations:

$$T = \sum_{n=1}^{16} \frac{y^i}{87} = 107.83$$

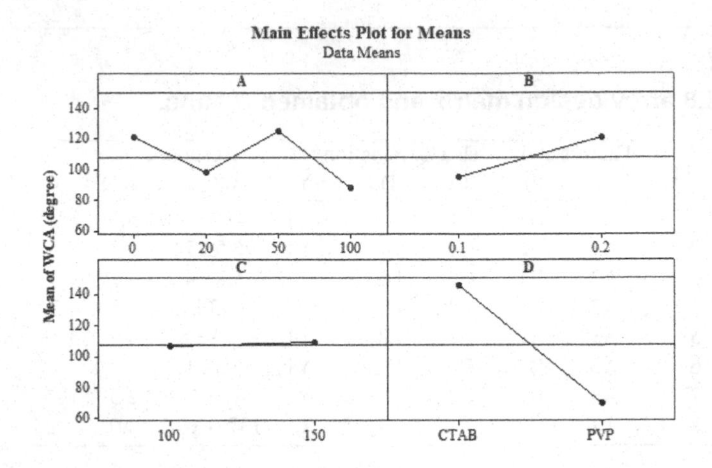

Figure 5.1 Factors main effect plot.

Table 5.4
Main effect of factors and their ranking.

Sample	Factors and Levels Combination			
	A	B	C	D
1	120.9	94.8	107.0	145.8
2	98.05	120.93	108.65	69.83
3	124.75			
4	87.65			
Delta	37.10	26.18	1.67	76.02
Rank	2	3	4	1

$$Y_{pred} = T + \sum_{i=1}^{5}(X_f - T)$$

$$Y_{pred} = T + (\overline{A}_3 - T) + (\overline{B}_2 - T) + (\overline{C}_2 - T) + (\overline{D}_1 - T)$$

where Y_{pred} is the predicted response at the optimum conditions, T is the grand average of the responses, and X_f is the total water contact angle for parameters at the optimum levels. Predicted response was at the optimum conditions 165.8±1.2° and CAH 1.5±0.5°, respectively, which showed an excellent superhydrophobicity. The obtained WCA confirmed that there is a strong agreement between the experimental (165.8°) and predicted values (176.68°). The small prediction error (6.5%) proved the reproducibility and high accuracy of the experimental results. The optical image of water droplet on the surface of the sample 9 is shown in Figure 5.2.

Table 5.5
ANOVA results.

Source of Variance	Sum Sq	Df	Mean sq	P-value	F Ratio	Contribution %
Seed Lauer Solution concentration (mM)	1920	3	640	0.026	7.29	12.9
Initial concentration of Zn (II) ions (M)	1370	1	0.015	7.65	9.2	
drying temperature	5.6	1	5.6	0.0227	3.65	0.03
Additive type	11559.6		11559.6	0.05		
Residual error	0.8		1	0.8		
total	14856	7				

Figure 5.2 Optical image of water droplet on the sample 9 (optimum sample).

5.5.2 XRD RESULTS

The XRD analysis was carried out to characterize the crystal structure and growth orientation related to the sample 9 Figure 5.3. The present characteristic peaks at 2 of 43.5, 50.7, and 74.6 are attributed to the cubic structure of 304-stainless steel substrate (card # 33-0397 of JCPDS) (4). The other diffraction peaks are labeled with the wurtzite-type structure of hexagonal ZnO according to the standard pattern of JCPDS card # 00-003-0888 (51). The sharpness and intensity of the peaks prove a well-crystalized ZnO formation on the stainless steel. In addition, the most intense peak at diffraction angle (2°) of 36.2 (the (101) plane) corresponds to the higher density of nanowires. Similar results are reported for preparation of ZnO thin films on the various substrates (26; 15).

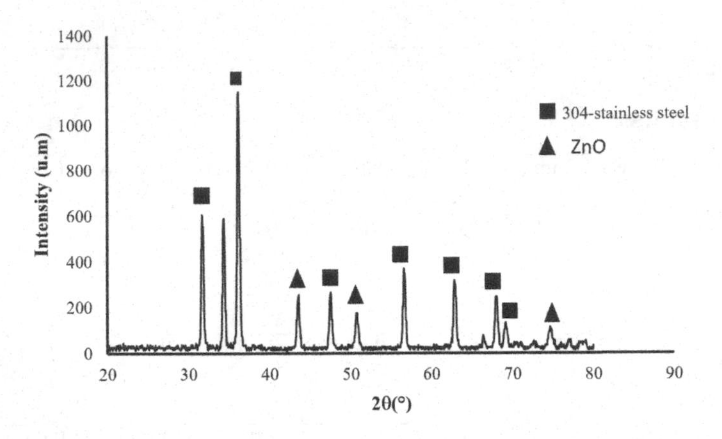

Figure 5.3 XRD pattern of sample 9 on the 304-stainless steel mesh.

5.5.3 SEM RESULTS

The surface morphology of fabricated ZnO film was investigated by scanning electron microscopy (SEM). The SEM images of the original uncoated stainless-steel mesh, ZnO seed layer, and hierarchical structured ZnO coated mesh prepared at the optimal conditions (sample 9) are shown in 5.4. The SEM top view of the original mesh shows quite a smooth surface with an average pore size of about 44 μm as a substrate in Figure 5.4 part a. Figures 5.4b and c present the SEM images of the seed layer. It can be seen that the ZnO seed is uniformly distributed on the substrate with an average diameter of 310 nm. After deposition of ZnO on the seed layer, the pore size of the mesh decreased significantly, as it is shown in Figure 5.4d. From the higher magnification SEM images of sample 9 (Figures 5.4e and f), it reveals the hierarchical ZnO micro/nanorods on the substrate are fabricated. The presence of micro/nano dual-scale roughness on the stainless-steel mesh is essential for the superhydrophobicity. Figures 5.4g and h show the SEM images of sample 6. This sample was prepared using PVP as an organic additive. As it can be observed, the aspect ratio (the mean length to diameter ratio) of ZnO rods is considerably reduced. The additives such as PVP and CTAB are non-polar chelating agents that can be attached to the non-polar facets of ZnO and promote the preferential orientation growth (50; 39; 31). In fact, the orientation control of ZnO nanowires is achieved using PVP and CTAB. It is noteworthy that the molecular weight of PVP is larger than CTAB, which can cap a large surface area, and the vertical growth of ZnO rods may be restricted effectively (39; 3). Consequently, rods with smoother rounded tips were formed on the stainless-steel mesh, as evidenced by SEM results (Figures 1.4g and h). In fact, the rate of nucleation is faster than that of ZnO rods growth using PVP as additive. In contrast, CTAB is a cationic agent with lower molecular weight which decreases the surface tension (34; 1). This phenomenon leads to form individual wire-like ZnO and formation of rough surface morphology, as shown in Figure 5.4e. In conclusion, the formation of smooth ZnO films on the substrate using PVP as an additive can result in increased hydrophilic properties of the surface.

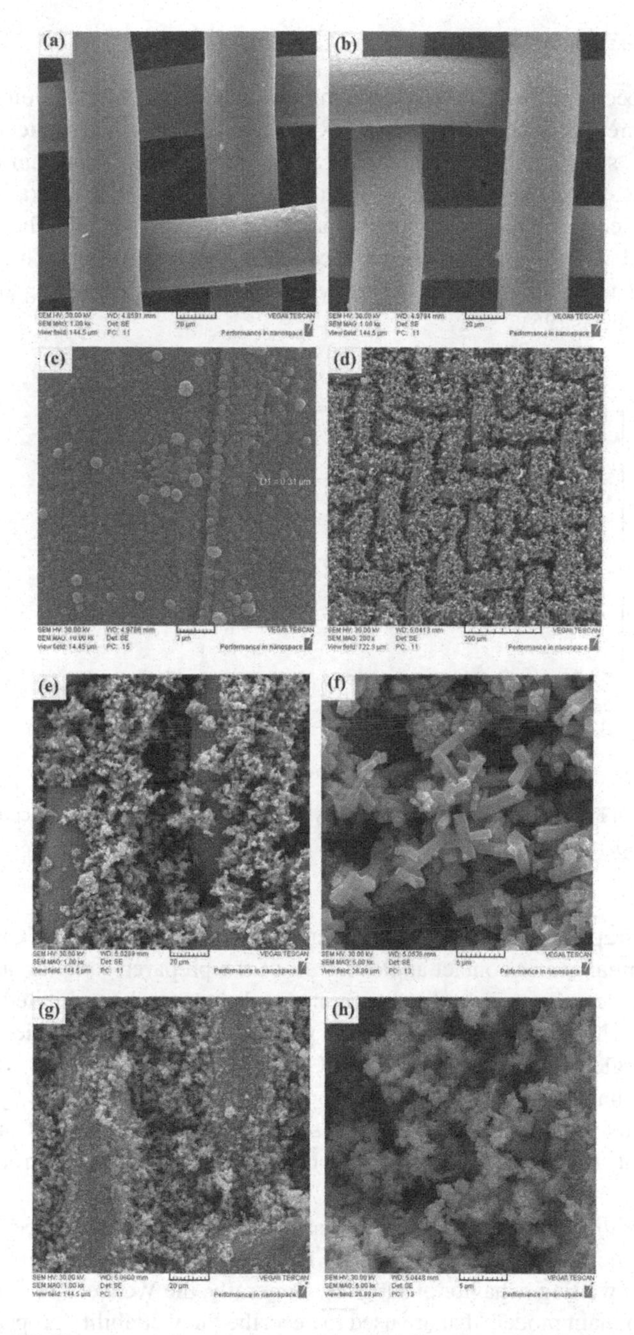

Figure 5.4 SEM images of: (a) the 304-stainless-steel mesh surface before the deposition stage, (b-c) ZnO seed layer, (d-f) ZnO surface deposited at the optimum condition, (g-h) sample 6.

5.5.4 ATR-FTIR ANALYSIS

ATR-FTIR spectroscopy was used to examine the chemical composition of the ZnO surfaces. Figure 5.5 displays the ATR-FTIR spectrum of ZnO rods scratched from the stainless-steel substrate. This sample was prepared under the optimum conditions. The sharp and strong band at 450 $cm^{-}1$ is related to the stretching vibration of Zn-O bonds, which confirms the purity of the prepared samples (9; 12). The weak extra peak observed at 1634 $cm^{-}1$ can be reduced to ZnO vibrations (24). The broad peak at 3400 $cm^{-}1$ may be ascribed to hydroxyl groups for both stretching and bending mode of vibrations (25).

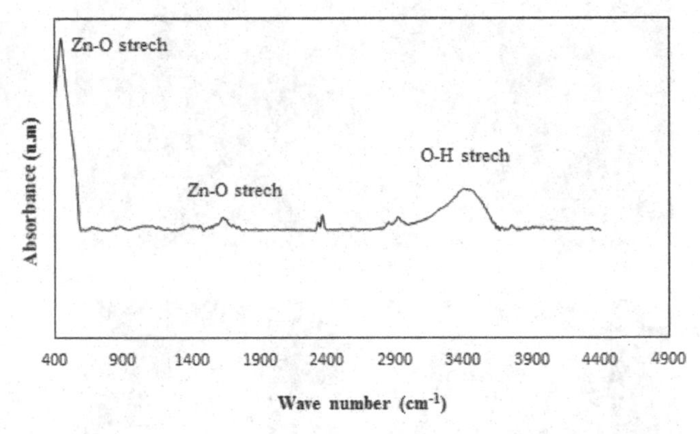

Figure 5.5 ATR-FTIR spectrum of ZnO rods was scratched from the surface of sample 9. Wettability properties.

The water-repellency of samples was evaluated by measuring the WCA and CAH values. The mean water contact angles for all of the prepared samples are shown in Figure 5.5. As can be seen in this figure, most of the samples exhibited hydrophobic properties. Moreover, samples 4 and 5 presented superhydrophobicity with water contact angles higher than 150°C and CAH values lower than 10. These results clearly show that the superhydrophobic properties can be achieved just by regulating the micro/nano-structured ZnO rods on the stainless steel mesh without any chemical post-treatment. Figure 5.7 indicates the optical images of the water droplets on the samples.

The wettability behavior of surfaces was determined by the surface free energy and topographical structure. Contact angle hysteresis is the important parameter for evaluating the wetting behavior of surfaces. Generally, the Wenzel and Cassie-Baxter states are two main models that are used to describe the wettability properties of surfaces (52). The important assumption of the Wenzel model is that the water droplet can fully penetrate to the rough features and adhere more strongly to the surface (32). In contrast, in the Cassie-Baxter state, the droplet suspends on the groove-texture of solid surfaces, and the air pocket entrapment on the solid-liquid contact occurs (54). This property leads to minimizing the water adhesion and decreasing the CAH value.

Table 5.6
f_1 and f_2 values for the super hydrophobic surfaces.

Sample	f_1	f_2
4	0.12	0.89
5	0.11	0.88
9	0.02	0.98

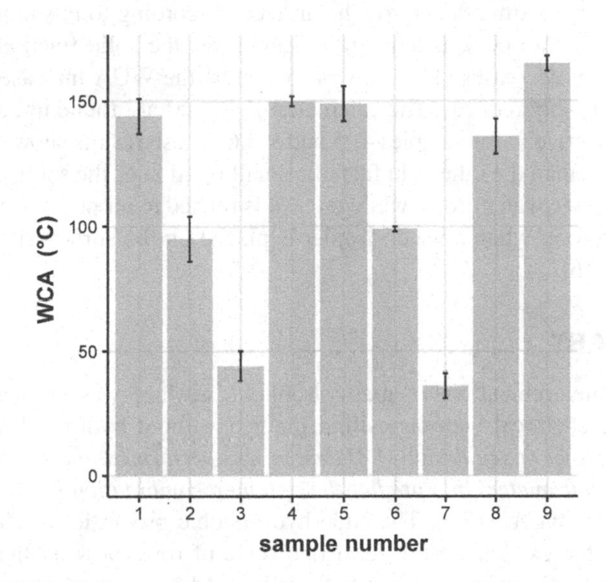

Figure 5.6 WCA of all prepared samples.

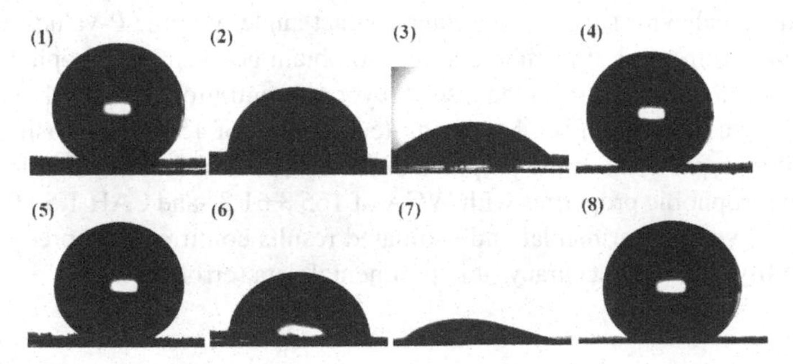

Figure 5.7 Optical images of water droplet on all samples (1-8).

So, the Wenzel model is usually applied to define the wettability properties of surfaces with CAH higher than 20.

Whereas, the Cassie state is predominant for the surfaces with CHA lower than 10 (13). Considering the ZnO rods on the stainless-steel mesh creating micro/nano dual-scale roughness, the water drops stay on the ZnO rods array and form a three-phase interface (solid-air-liquid), which leads to higher WCA and lower CAH. Therefore, the wetting behavior of surfaces follows the Cassie state theory described by Eq. 5.1.

$$\cos\theta^* = f_1\cos\theta - f_2(f_1 + f_2 = 1) \tag{5.1}$$

In this equation, f_1 and f_2 express the fractions of the solid surface which have contact with water droplet and air, respectively. θ^* and θ and are the apparent water contact angles for the smooth and rough surfaces. According to this model, with an increase in the air trap pockets among the ZnO rods, the solid fraction contacting with water droplet decreases (11; 43). Consequently, the WCA increases and water adhesion considerably reduces. The calculated f_1 values are found to be 0.12, 0.11, 0.02, and 5 respectively, for samples 4, 5 and 9 5.6. These results show f_1 value for sample 9 is lower than the others. In fact, the small fraction of the solid surface (2%) is in contact with the water drops, while the rest is related to air pockets (98%) in this condition. Therefore, when a water droplet is placed on the surface, it can quickly roll off (23; 32; 16).

5.6 SUMMARY

A micro/nano hierarchical ZnO superhydrophobic surface was fabricated via the simple and fast chemical bath deposition method without hydrophobization post-treatment. *A Taguchi mixed design (L8) design was performed to evaluate the effective controlling parameters and predict the optimum combination levels and factors that lead to more WCA values.* The superhydrophobic properties of ZnO surfaces were related to the existence of at least one scale of roughness on the substrates, and this characteristic is improved if both micro and nano-roughness are applied. In the parameter range considered, the additive type, seed layer solution concentration, and initial concentration of Zn (II) ions result to be most effective parameters on the wetting behavior surface and water contact angle with the P-values less than 0.05. The optimum level of parameters used to obtain good superhydrophobic characteristics of ZnO surfaces included a seed layer concentration of 50 mM, a zinc (II) ions initial concentration of 0.2 M, drying temperature of 150°C, and using CTAB as the additive type. The sample prepared at the optimal conditions exhibited excellent superhydrophobic properties with WCA of 165.8±1.2° and CAH 1.5±0.5°. The difference between experimental and estimated results confirmed the predictability, reproducibility, and high accuracy of experimental data (error 6.5%).

Bibliography

1. K. Akhil, J. Jayakumar, G. Gayathri, and S. Sudheer Khan. Effect of various capping agents on photocatalytic, antibacterial and antibiofilm activities of ZnO nanoparticles. *Journal of Photochemistry and Photobiology B: Biology*, 160:32–42, jul 2016.
2. Anam Asghar, Abdul Aziz Abdul Raman, and Wan Mohd Ashri Wan Daud. A Comparison of Central Composite Design and Taguchi Method for Optimizing Fenton Process. *The Scientific World Journal*, 2014:1–14, 2014.
3. Sanjaya Brahma, P. Jaiswal, K. S. Suresh, Kuang Yao Lo, Satyam Suwas, and S. A. Shivashankar. Effect of substrates and surfactants over the evolution of crystallographic texture of nanostructured ZnO thin films deposited through microwave irradiation. *Thin Solid Films*, 593:81–90, oct 2015.
4. Yongjun Chen, Zhangfa Tong, and Lijic Luo. Boron Nitride Nanowires Produced on Commercial Stainless Steel foil. *Chinese Journal of Chemical Engineering*, 16(3):485–487, jun 2008.
5. Er Chieh Cho, Cai Wan Chang-Jian, Hsin Chou Chen, Kao Shuh Chuang, Jia Huei Zheng, Yu Sheng Hsiao, Kuen Chan Lee, and Jen Hsien Huang. Robust multifunctional superhydrophobic coatings with enhanced water/oil separation, self-cleaning, anti-corrosion, and anti-biological adhesion. *Chemical Engineering Journal*, 314:347–357, 2017.
6. Ying Tao Chung, Muneer M. Ba-Abbad, Abdul Wahab Mohammad, Nur Hanis Hayati Hairom, and Abdelbaki Benamor. Synthesis of minimal-size ZnO nanoparticles through sol-gel method: Taguchi design optimisation. *Materials and Design*, 87:780–787, dec 2015.
7. Marcelo Azevedo Costa, Antônio Pádua Braga, and Benjamin Rodrigues De Menezes. Improving neural networks generalization with new constructive and pruning methods. *Journal of Intelligent and Fuzzy Systems*, 2003.
8. Russell J. Crawford and Elena P. Ivanova. *Superhydrophobic Surfaces*. Elsevier Ltd, feb 2015.
9. Debadrito Das, Animesh Kumar Datta, Divya Vishambhar Kumbhakar, Bapi Ghosh, Ankita Pramanik, and Sudha Gupta. Conditional optimisation of wet chemical synthesis for pioneered ZnO nanostructures. *Nano-Structures and Nano-Objects*, 9:26–30, feb 2017.
10. Haihui Di, Zongxue Yu, Yu Ma, Fei Li, Liang Lv, Yang Pan, Yan Lin, Yi Liu, and Yi He. Graphene oxide decorated with Fe3O4 nanoparticles with advanced anticorrosive properties of epoxy coatings. *Journal of the Taiwan Institute of Chemical Engineers*, 64:244–251, jul 2016.
11. Xin Du, Xing Huang, Xiaoyu Li, Xiangmin Meng, Lin Yao, Junhui He, Hongwei Huang, and Xueji Zhang. Wettability behavior of special microscale ZnO nail-coated mesh films for oil-water separation. *Journal of Colloid and Interface Science*, 458:79–86, nov 2015.

12. Shuwang Duo, Yangyang Li, Zhao Liu, Ruifang Zhong, Tingzhi Liu, and Haiming Xu. Preparation of ZnO from 2 D nanosheets to diverse 1 D nanorods and their structure, surface area, photocurrent, optical and photocatalytic properties by simple hydrothermal synthesis. *Journal of Alloys and Compounds*, 695:2563–2579, feb 2017.
13. Houda Ennaceri, Lan Wang, Darja Erfurt, Wiebke Riedel, Gauri Mangalgiri, Asmae Khaldoun, Abdallah El Kenz, Abdelilah Benyoussef, and Ahmed Ennaoui. Water-resistant surfaces using zinc oxide structured nanorod arrays with switchable wetting property. *Surface and Coatings Technology*, 299:169–176, aug 2016.
14. S. Esmailzadeh, S. Khorsand, K. Raeissi, and F. Ashrafizadeh. Microstructural evolution and corrosion resistance of super-hydrophobic electrodeposited nickel films. *Surface and Coatings Technology*, 283:337–346, dec 2015.
15. Wei Feng, Jie Chen, and Chun yan Hou. Growth and characterization of ZnO needles. *Applied Nanoscience (Switzerland)*, 4(1):15–18, jan 2014.
16. Aziz Fihri, Enrico Bovero, Abdullah Al-Shahrani, Abdullah Al-Ghamdi, and Gasan Alabedi. Recent progress in superhydrophobic coatings used for steel protection: A review, may 2017.
17. X. K. Gao, Q. H. Liu, and Z. J. Liu. Robust design of head interconnect for hard disk drive. In *Journal of Applied Physics*, volume 97, may 2005.
18. H. Ghayour, H. R. Rezaie, Sh Mirdamadi, and A. A. Nourbakhsh. The effect of seed layer thickness on alignment and morphology of ZnO nanorods. *Vacuum*, 86(1):101–105, jul 2011.
19. Simone Ghio, Giovanni Paternoster, Ruben Bartali, Pierluigi Belluti, Maurizio Boscardin, and Nicola M. Pugno. Fast and large area fabrication of hierarchical bioinspired superhydrophobic silicon surfaces. *Journal of the European Ceramic Society*, 36(9):2363–2369, aug 2016.
20. M. Giahi and S. Toutounchi. Synthesis of ZnO nanoparticles and its application in photocatalytic degradation of LABS by the trial-and-error and Taguchi methods. *Russian Journal of Applied Chemistry*, 89(5):823–829, may 2016.
21. Tao Hang, Anmin Hu, Huiqin Ling, Ming Li, and Dali Mao. Super-hydrophobic nickel films with micro-nano hierarchical structure prepared by electrodeposition. *Applied Surface Science*, 256(8):2400–2404, feb 2010.
22. Maryam Khosravi and Saeid Azizian. Preparation of superhydrophobic and superoleophilic nanostructured layer on steel mesh for oil-water separation. *Separation and Purification Technology*, 172:366–373, jan 2017.
23. Min Hwan Kim, Hisuk Kim, Kwan Soo Lee, and Dong Rip Kim. Frosting characteristics on hydrophobic and superhydrophobic surfaces: A review, 2017.
24. Harish Kumar and Renu Rani. Structural and Optical Characterization of ZnO Nanoparticles Synthesized by Microemulsion Route. *International Letters of Chemistry, Physics and Astronomy*, 19:26–36, oct 2013.
25. Marco Laurenti, Valentina Cauda, Rossana Gazia, Marco Fontana, Vivian Farías Rivera, Stefano Bianco, and Giancarlo Canavese. Wettability Control on ZnO Nanowires Driven by Seed Layer Properties. *European Journal of Inorganic Chemistry*, 2013(14):2520–2527, may 2013.

26. Geun Hyoung Lee. Relationship between crystal structure and photoluminescence properties of ZnO films formed by oxidation of metallic Zn. *Electronic Materials Letters*, 6(4):155–159, dec 2010.
27. Hong Li, Yushan Li, and Qinzhuang Liu. ZnO nanorod array-coated mesh film for the separation of water and oil. *Nanoscale Research Letters*, 8(1):1–6, 2013.
28. Jian Li, Long Yan, Yuzhu Zhao, Fei Zha, Qingtao Wang, and Ziqiang Lei. One-step fabrication of robust fabrics with both-faced superhydrophobicity for the separation and capture of oil from water. *Physical Chemistry Chemical Physics*, 17(9):6451–6457, mar 2015.
29. J. Lomga, P. Varshney, D. Nanda, M. Satapathy, S. S. Mohapatra, and A. Kumar. Fabrication of durable and regenerable superhydrophobic coatings with excellent self-cleaning and anti-fogging properties for aluminium surfaces. *Journal of Alloys and Compounds*, 702:161–170, 2017.
30. Minglin Ma and Randal M. Hill. Superhydrophobic surfaces, oct 2006.
31. C. M. Mbulanga, Z. N. Urgessa, S. R. Tankio Djiokap, J. R. Botha, M. M. Duvenhage, and H. C. Swart. Surface characterization of ZnO nanorods grown by chemical bath deposition. *Physica B: Condensed Matter*, 480:42–47, jan 2016.
32. Athanasios Milionis, Eric Loth, and Ilker S. Bayer. Recent advances in the mechanical durability of superhydrophobic materials, mar 2016.
33. Adel M.A. Mohamed, Reza Jafari, and Masoud Farzaneh. An optimization of superhydrophobic polyvinylidene fluoride/zinc oxide materials using Taguchi method. *Applied Surface Science*, 288:229–237, jan 2014.
34. Guru Nisha Narayanan and A. Karthigeyan. Influence of different concentrations of Cetyltrimethylammonium bromide on morphological, Structural and optical properties of Zinc Oxide nanorods. In *Materials Today: Proceedings*, volume 3, pages 1762–1767. Elsevier Ltd, 2016.
35. Navdeep Pandey, K. Murugesan, and H. R. Thomas. Optimization of ground heat exchangers for space heating and cooling applications using Taguchi method and utility concept. *Applied Energy*, 190:421–438, 2017.
36. R Panneerselvam. *DESIGN AND ANALYSIS OF EXPERIMENTS*. PHI, 2012.
37. Thang Pham, Anna P. Goldstein, James P. Lewicki, Sergei O. Kucheyev, Cheng Wang, Thomas P. Russell, Marcus A. Worsley, Leta Woo, William Mickelson, and Alex Zettl. Nanoscale structure and superhydrophobicity of sp2-bonded boron nitride aerogels. *Nanoscale*, 7(23):10449–10458, jun 2015.
38. Sareh Poorebrahimi and Reza Norouzbeigi. A facile solution-immersion process for the fabrication of superhydrophobic gibbsite films with a binary micro-nano structure: Effective factors optimization via Taguchi method. *Applied Surface Science*, 356:157–166, nov 2015.
39. M.Y.A. Rahman, A.A. Umar, R. Taslim, and M.M. Salleh. Effect of surfactant on the physical properties of ZnO nanorods and the performance of ZnO photoelectrochemical cell. *Journal of Experimental Nanoscience*, 10(8):599–609, may 2015.
40. Ranjit K. Roy. *A Primer on the Taguchi Method, Second Edition*. Society of Manufacturing Engineers, 2010.
41. S. Santangelo, G. Messina, A. Malara, N. Lisi, T. Dikonimos, A. Capasso, L. Ortolani, V. Morandi, and G. Faggio. Taguchi optimized synthesis of graphene films by copper catalyzed ethanol decomposition. *Diamond and Related Materials*, 41:73–78, jan 2014.

42. Vincent Senez, Vincent Thomy, and Renaud Dufour. *Nanotechnologies for Synthetic Super Non-wetting Surfaces*, volume 9781848215795. Wiley Blackwell, sep 2014.
43. Ashoka Siddaramanna, N. Saleema, and D. K. Sarkar. A versatile cost-effective and one step process to engineer ZnO superhydrophobic surfaces on Al substrate. *Applied Surface Science*, 311:182–188, aug 2014.
44. Hao Jie Song, Xiang Qian Shen, Hai Yan Ji, and Xiao Jing Jing. Superhydrophobic iron material surface with flower-like structures obtained by a facile self-assembled monolayer. *Applied Physics A: Materials Science and Processing*, 99(3):685–689, jun 2010.
45. Ziqi Sun, Ting Liao, Kesong Liu, Lei Jiang, Jung Ho Kim, and Shi Xue Dou. Fly-Eye Inspired Superhydrophobic Anti-Fogging Inorganic Nanostructures. *Small*, 10(15):3001–3006, aug 2014.
46. Genichi Taguchi and Shih Chung Tsai. Quality Engineering (Taguchi Methods) For The Development Of Electronic Circuit Technology. *IEEE Transactions on Reliability*, 44(2):225–229, 1995.
47. Dongliang Tian, Xiaofang Zhang, Xiao Wang, Jin Zhai, and Lei Jiang. Micro/nanoscale hierarchical structured ZnO mesh film for separation of water and oil. *Physical Chemistry Chemical Physics*, 13(32):14606–14610, aug 2011.
48. M. A. Titu and A. B. Pop. Contribution on Taguchi's Method Application on the Surface Roughness Analysis in End Milling Process on 7136 Aluminium Alloy. In *IOP Conference Series: Materials Science and Engineering*, 2016.
49. N. Valipour M., F. Ch Birjandi, and J. Sargolzaei. Super-non-wettable surfaces: A review, apr 2014.
50. Tan T. Vu, Laura Del Río, Teresa Valdés-Solís, and Gregorio Marbán. Tailoring the synthesis of stainless steel wire mesh-supported ZnO. *Materials Research Bulletin*, 47(6):1577–1586, jun 2012.
51. Huaiyuan Wang, Yixing Zhu, Ziyi Hu, Xiguang Zhang, Shiqi Wu, Rui Wang, and Yanji Zhu. A novel electrodeposition route for fabrication of the superhydrophobic surface with unique self-cleaning, mechanical abrasion and corrosion resistance properties. *Chemical Engineering Journal*, 303:37–47, nov 2016.
52. Mingshan Xue, Wenfeng Wang, Fajun Wang, Junfei Ou, and Wen Li. Design and understanding of superhydrophobic ZnO nanorod arrays with controllable water adhesion. *Surface and Coatings Technology*, 258:200–205, nov 2014.
53. Dongyun Yu and Jintao Tian. Superhydrophobicity: Is it really better than hydrophobicity on anti-corrosion? *Colloids and Surfaces A: Physicochemical and Engineering Aspects*, 445:75–78, mar 2014.
54. Tao Zhang, Linda Y.L. Wu, and Zhenfeng Wang. Smart UV/Visible light responsive polymer surface switching reversibly between superhydrophobic and superhydrophilic. *Surface and Coatings Technology*, 320:304–310, jun 2017.
55. Jing Hui Zhi, Li Zhi Zhang, Yuying Yan, and Jie Zhu. Mechanical durability of superhydrophobic surfaces: The role of surface modification technologies. *Applied Surface Science*, 392:286–296, jan 2017.
56. Hai Zhu and Zhiguang Guo. Understanding the Separations of Oil/Water Mixtures from Immiscible to Emulsions on Super-wettable Surfaces. *Journal of Bionic Engineering*, 13(1):1–29, jan 2016.

5.7 AN EXAMPLE OF EVOLUTIONARY DESIGN OF EXPERIMENT: PREDICTION OF THE AGING OF POLYMERS

Oxana Ye. Rodionova[2] and Alexey L. Pomerantsev[2]

5.8 INTRODUCTION

Properties of a polymer are not defined by its chemical formula. Molecules of the most simple (linear) polymer have different lengths, and they can interact forming complex structures. This leads to the irreproducibility of material properties. The same material made in different conditions demonstrates completely different properties. In experiments we observe large property variations, which consist of a small measurement error together with a huge random variation related to the diversity of the samples. Polymer systems are far from the thermodynamic equilibrium. Their structure and therefore their properties are changing with time even in the absence of the external influence. Moreover, many polymers are very sensitive to the environment. Oxygen, light, and temperature affect the polymer properties greatly. For example, polyethylene protected from light keeps the initial mechanical properties for a long time (more than 30 years). The same polyethylene exposed to light loses these properties in 2 years. A complex physicochemical process of changing a polymer's properties with time is called "the aging of polymers". Very often the lifespan of polymers is more than ten years, therefore the statistics of failures is unavailable. For these reasons the prediction of polymer aging is considered to be the most difficult problem in modelling of the material behavior (1).

To solve this problem we have to develop a physical-chemical model, which describes the aging process over time within a range of normal or adverse conditions. This model is calibrated using the results obtained during the accelerated aging tests (AAT), and then predicted to the normal operation conditions. During AAT the polymer samples are exposed to the aging at the conditions (factors) that are more severe than the normal ones to speed up the regular aging processes. The properties of interest y are measured in the course of AAT and the obtained results are used to develop a model that describes aging over time t, regarding factors X:

$$y = f(t, \mathbf{X}, \mathbf{a}) \tag{5.2}$$

that describes aging over time t regarding factors $\mathbf{X}$. The model depends on the unknown parameters $\mathbf{a}$, which are to be estimated using the Least Squares Method as a minimum of the objective function:

$$S(a) = \sum (f(t_i, \mathbf{X_i}, \mathbf{a}) - y_i)^2 \tag{5.3}$$

[2]Semenov Federal Research Center for Chemical Physics RAS, Kosygin str. 4, 119991, Moscow, Russia, rcs@chph.ras.ru

After assessing model parameters, we can extrapolate the model to the natural (operation) conditions and obtain the material properties **y**. In case the critical levels of *y*'s are known, it is possible to calculate the lifespan at which the levels are attained. This is a general scheme of the polymer aging prediction, but there are several difficult issues that merit special attention.

AATs provide the main data source used for the evaluation of polymer stability, for the prediction of its behavior at various conditions, and for the comparison of different polymer compositions, formulations, additives, etc. The quality of this analysis greatly depends on the quality of the obtained AAT data. A proper duration of AAT provides us with a complete information regarding a whole lifespan of a polymer up to its failure, and, in this way, gives a possibility to develop the model of aging presented in Eq. 5.2. From the theory of polymer aging, it is known that the rate of a chemical reaction depends on the conditions in which the reaction proceeds. In particular, the dependence of a rate constant k is presented by the Arrhenius equation:

$$k = k_0 exp\left(-\frac{E}{RT}\right) \tag{5.4}$$

where T is temperature (K), k_0 is an entropic factor (pre-exponential), E is the activation energy, and R is the gas constant. From this equation, it follows that the aging process accelerates when temperature increases. This fact gives a basis for AAT and prescribes two parameters that can be managed in designing of the AAT experiments. They are time t, and temperature, T.

In the course of AAT, it is important to meet two main conditions, only under which the extrapolation is possible. First of all we have to achieve an appropriate aging depth in a short time. This depth should be about the critical value that characterizes a material failure. This can be done by increasing the testing temperature. On the other hand, during AAT, we have to retain the mechanism of the natural aging of a polymer. For example, in AAT of polyethylene, temperature should be lower than the polymer melting point. This condition forces us to decrease the testing temperature. Thus, the objective of the design of experiments (DOE) for AAT procedure is to find the balance between the shorter time t, and the lower temperature T.

Aging is characterized by the simultaneous changes of the various material properties. The most important of them are controlled in the course of AAT. Each property is described by its own model. Thus, in modelling we have to optimize not single, but several objective functions. However, the same aging process underlies these changes. That is why, common parameters are participating in the mathematical models of various properties. Such parameters as pre-exponential k_0 and the activation energy E, presented in Eq. 5.4, are common for the different models, and these parameters cannot be found by minimizing each objective function separately. Conventional classical approach prescribes building of a multi-response regression for simultaneous analysis of experimental data, in which each model comes with a specified weight value due to different errors in different responses. However, those errors are, in general, unknown, so elaboration of this general optimization becomes a rather complicated iteration procedure. To resolve this problem, the method of Successive Bayesian Estimation (SBE) has been developed (3). This technique allows processing of data successively for every response. The SBE method has been widely used for the data modelling (6; 7; 4; 2). In this research, SBE is applied for prediction of the rubber lifespan.

The biggest challenge is the prediction problem, which means extrapolation of a model over the factors that are beyond the observed area. Conventional validation methods, such as the test or cross validation, are not applicable as they are intended for interpolation only. Strictly speaking, there are no mathematical/statistical procedures, which can guarantee the prediction. On the other hand, a 'hard' physicochemical modelling cannot assure the results either. Any complication of a model, which describes the time limited experiment, leads to overfitting, multicollinearity, and, as result, to unreliable prediction (5). Polymer aging is a complex physicochemical process that is characterized by numerous parameters, which can be ranged in the order that they affect the aging process. In very many practical cases, aging is determined by one or two, rarely by three, main parameters. To make a right choice it is necessary to fulfil two main conditions:

- Parameters under consideration should be common for all individual models;
- Influence of these parameters on the aging should be appraised of in the course of AAT

5.8.1 EVOLUTIONARY DESIGN OF EXPERIMENT FOR ACCELERATED AGING TESTS

We demonstrate the AAT design using the tire rubber as the object. A typical rubber formulation includes- natural rubber, synthetic rubber, and sulfur. These components are vulcanized (cured) at the curing temperature T_{cur}, during the curing time t_{cur} to produce a durable material by forming cross-links (bridges) between individual polymer chains.

We consider that elongation at break, $\overline{\varepsilon}$, is the most sensitive property, because this characteristic first and foremost reflects the changes in structural homogeneity and defectiveness which occur during the aging of a rubber. Therefore we use this characteristic as the basic characteristic in the special procedure for evolutionary design of experiment (EDOE). EDOE assists researchers in avoiding extra expenditure of time and labor, and in obtaining reliable data in reasonable terms (1).

The procedure of EDOE consists of three stages- preliminary designing, correction step (steps), and final designing. At the first step, we perform AAT at the highest possible temperature T_{max} during two minimal time periods. Temperature T_{max} is chosen about 15-30 degrees less than curing temperature T_{cur}. Two measurements of ELB should be performed at two time periods- t_1 and $t_2 = 2t_1$. Value t_1 is calculated using the Arrhenius law:

$$t_1 = t_{cur} exp \left[\frac{E}{R} \left(\frac{1}{T_{max}} - \frac{1}{T_{cur}} \right) \right] \tag{5.5}$$

where t_{cur} is the curing time, and T_{cur} is the curing temperature. Activation energy E is chosen regarding the rubber recipe ($E/R = 10,000$ for the sulfur curing agents, and 12,500 for the compounds that do not have free sulfur). Our goal is to design the AAT experiments that provide the achievement of a given aging depth, which can be presented as a relative change of ELB, i.e.,

$$\Delta\overline{\varepsilon} = 1 - \overline{\varepsilon}(t)/\overline{\varepsilon}(0) \quad (5.6)$$

where $\overline{\varepsilon}(0)$ is the initial ELB value measured before aging.

The results of measurements, including values of $\overline{\varepsilon}(0)$, are used to develop the AAT design at any temperature. First of all we should analyze obtained data $\overline{\varepsilon}(0)$, $\overline{\varepsilon}(t_1)$, and $\overline{\varepsilon}(t_2)$ to be sure that ELB essentially decreases. As far as we use a heuristic formula for determination of t_1, there is a chance that an appropriate aging depth will not be achieved during the preliminary experiments. Taking into account that a general error of ELB measurement is about 20-30%, we should expect $\Delta\overline{\varepsilon}$ not less than 0.25 in the preliminary testing. In another case, the result of the time extrapolation may be strongly overestimated. Moreover, sometimes a small increase of ELB is observed in the initial stage of rubber thermo-aging. To avoid such faults, we use the student test to verify that ELB actually decreases during the preliminary tests. In case ELB does not go down essentially, we conduct an additional test at time $t_3 = 3t_1$, etc.

The procedure assumes not more than three corrections of the time points, which are calculated as $t_m = mt_1$. In the result, we obtain three values of $\overline{\varepsilon}(t)$ at three time-points t_{m-2}, t_{m-1}, t_m. This data is analyzed using a simplified model for ELB:

$$\Delta\overline{\varepsilon} = \varepsilon(0)exp(-kt) \quad (5.7)$$

where $\varepsilon(0)$ is the initial value and k is the rate parameter. These parameters are estimated by the Least Squares Method that is applied to the logarithmic transformed model:

$$y(t) = b - kt \quad (5.8)$$

where $y = ln\varepsilon$ and $b = ln\varepsilon_0$. Logarithmic transformation is valid because we use only the initial part of the ELB curve, where error distortion is not substantial. The quality of estimation is characterized by the variance-covariance matrix $\mathbf{C} = cov(b,k)$ calculated in the usual way.

Now, we can calculate t_{max} that is the duration of aging at temperature T_{max}, meaning that a given aging depth ε_{des} should be achieved. In general, the aging model is more complex than its simplified version given in Eq. 5.6.

In particular, the rate of ELB decrease is notably smaller at the large depths of rubber aging. To account for these effects, we determine time tmax as the upper confidence bound, which is calculated as the positive root to the quadratic equation:

$$(k^2 - x_p^2 C_{kk})^2 t^2 + 2(bk + zk + x_p^2 C_{kb})t + (z^2 + b^2 - 2zb - x_p^2 C_{bb}) = 0 \quad (5.9)$$

where $z = ln(\varepsilon_{des})$, x_P is the P-quantile of the normal distribution, C_{xy} are the components of variance-covariance matrix $\mathbf{C}$, and P = 0.95.

The example is shown in Figure 5.8. Several control points (open dots), corresponding to period of six and nine hours, are measured after the EDOE data points

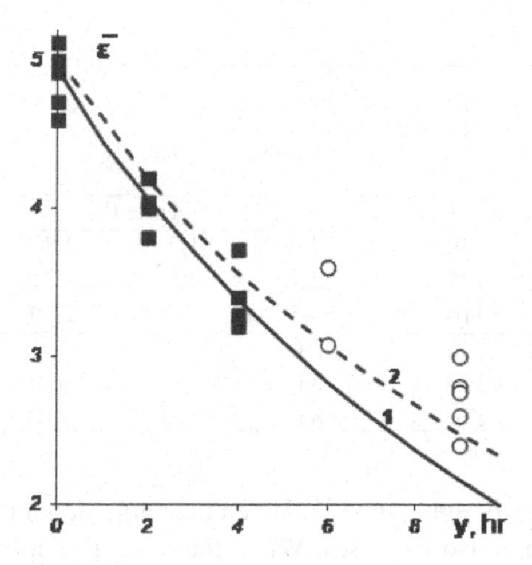

Figure 5.8 ELB model extrapolated over time. square EDOE data, circle control data, 1 mean curve, 2 0.99 confidence curve.

(black squares) are taken. It is evident that the exponential curve (1) goes apart from the control results, but the one-side confidence interval (2) corrects these deviations.

Usually, the rubber thermo-aging complies with the Arrhenius law. However, the activation energy may be different for the different compositions, e.g., E/R is varied from 10,000 till 12,500. To account for these variations we use an expression, in which the activation energy depends on the temperature extrapolation interval $\Delta T = T_{max}$–T:

$$E(T) = 10^3[1 + \alpha(1 - exp(-\beta\Delta T)] \tag{5.10}$$

Parameters $\alpha = 0.3$, $\beta = 0.06$ are chosen heuristically with the purpose to achieve a given depth of aging, i.e., overestimation is considered more preferable, than underestimation. Now we can compose the plan of experiments for any temperature T as:

$$t_{max} = t_{max}(T_{max})exp\left[\frac{E(T)}{R}\left(\frac{1}{T} - \frac{1}{T_{max}}\right)\right] \tag{5.11}$$

where $E(T)$ is calculated by 5.10. The period of AAT, $t_{max}(T)$, is divided into several time points in the following way. The first interval is a little longer than time of possible increasing of ELB:

$$t_1 = t_{cur}exp\left[\frac{E}{R}\left(\frac{1}{T} - \frac{1}{T_{cur}}\right)\right] \tag{5.12}$$

and other time points t_k, k=2,..,n, are calculated as $t_k = t_1 + k(t_{max}–t_1)/n$. The EDOE procedure is based on the half-empirical models. Therefore, the AAT design should be considered as a reasonable approximation, which requires some validation. In paper (1), a good conformity between the EDOE prediction and the results obtained in the real AAT was demonstrated for seven different rubber compounds. In the temperature interval $[T_{max}$–30, $T_{max}]$ and for the aging depth $\Delta\bar{\varepsilon} = 0.5$, the EDOE procedure

Table 5.7
Design of the AAT.

	Temperature (°C)	Time (h)			
EDOE	140	0.9	1.7	2.8	3.9
	125	2.1	4.1	7.7	12.0
	110	5.5	11.0	27.0	42.0
Experiment	140	2.0	4.0	6.0	9.0
	125	3.0	6.0	14.0	23.0
	110	8.0	16.0	38.0	60.0

provides a correct time of aging in 90%. When the aging depth is increased, the risk of underestimated time also increases. When the testing temperature is decreased, the risk of overestimated time increases.

5.9 PREDICTION OF RUBBER AGING BY ACCELERATED AGING TESTS

We demonstrate the capabilities of EDOE for the prediction of aging of the tire rubber (9). All samples are rectangular plaques of 200 mm $\times$ 200 mm $\times$ 2 mm, which were produced especially for this experiment. Tread tire rubber is formulated as follows- natural rubber, butadiene-styrene rubber, sulfur. The curing temperature is $T_{cur} = 160°C$ and curing time is t_{cur} =16 min. Five standard samples (ASTM D412-87) were prepared from each plaque for one test. In the course of AAT the following properties are measured using tensile machine Instron.

- $\overline{\varepsilon}$, is the elongation at break (ELB)
- $\overline{\sigma}$, is the ultimate tensile strength (UTS)
- λ_k, is the Young's moduli obtained at elongation k=1,2.3,4,5

Time is measured in hours, temperature in centigrade. In this case the AAT design was developed for three constant temperatures. Table 5.7 shows the design suggested by EDOE and the real AAT conditions used in the experiment.

As we mentioned before, the EDOE procedure is based on half-empirical models and the AAT design is of course approximate. The researcher should consider the EDOE results merely as an advise, but not as a strict instruction. In our case the periods of aging calculated by EDOE turned out to be less than the acceptable time of experiment. That is why we decided to prolong the experiment in line with the admissible period of AAT. However, the results of AAT (see Figure 5.9) show that the predefined aging depth has been achieved at an even shorter time of aging than was suggested by EDOE.

Apparently the deformation curve, i.e., a dependence of strain σ versus elongation ε, obtained for each sample, is the best data-set for the prediction of the mechanical properties of rubbers. The advantage of this approach is in reconstructing the "aging

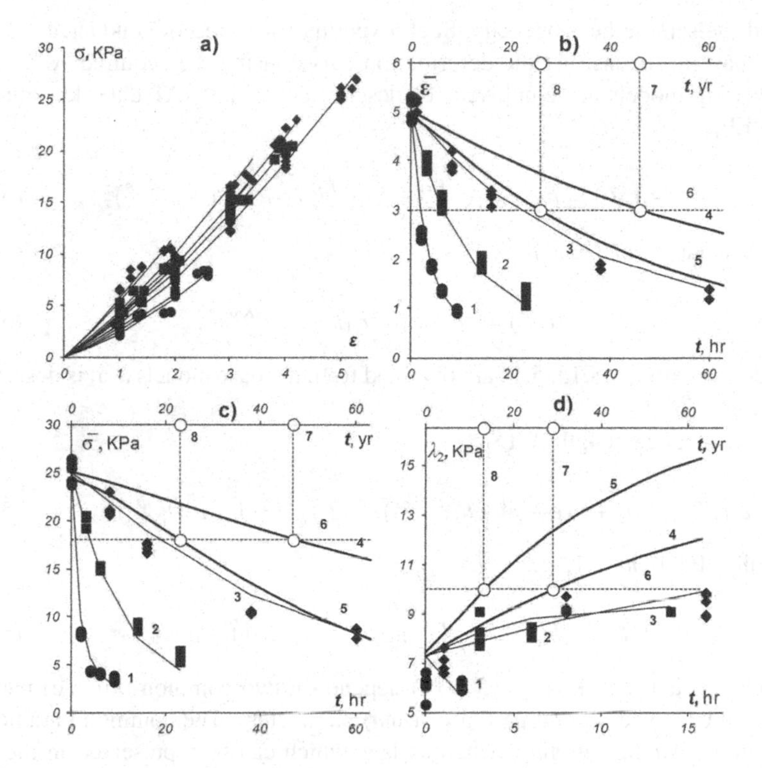

Figure 5.9 AAT data and the results of prediction to temperature T = 20(°C)- (a) deformation curves, (b) elongation at break, c) ultimate tensile strength, (d) second modulus. 1 (black circle) T = 140(°C), 2 (black square) T = 125(°C), 3 (white square) T = 110(°C), 4 predicted value, 5 0.95 confidence limit, 6 critical level, 7 mean time, 8 confidence time.

trajectory" that reflects the relationship between different properties in the course of aging in a compact form (9). The existence of a universal trajectory that is independent of the testing temperature confirms that all changes in the properties of a material are caused by a common aging process.

Let us suppose that we have a model for the deformation curve $\sigma(\varepsilon,t,T)$, which is valid for any elongation value ε, time t, and temperature T. From this model we can derive models for each modulus, for example,

$$\lambda_1(t,T) = \sigma(1,t,T) \tag{5.13}$$

In case we construct a model for the elongation at break $\varepsilon(t,T)$, we can derive a model for the ultimate tensile strength by combining two models as:

$$\overline{\sigma}(t,T) = \sigma(\overline{\varepsilon}(t,T),t,T) \tag{5.14}$$

Furthermore, if we have the deformation curves for each sample, we can determine the differences between the samples, and, as a result, detach the measurement error and the sample heterogeneity. Such experiments were not conducted. Therefore, we have to reconstruct the deformation curves using the acquired AAT data. The following models are employed for description of the AAT data deformation curve (DEF):

$$\sigma(\varepsilon,t,T) = b_0 + (b_1 e^{-K_1 t} + b_2 e^{-K_2 t})\varepsilon + b_3(1 - e^{-b_4 \varepsilon}) \tag{5.15}$$

Elongation at break (ELB):

$$\overline{\varepsilon}(t,T) = a_0 + a_1 e^{-} K_1 t + a_2 e^{-K_2 t} \tag{5.16}$$

The derived models 5.13, 5.14 are obtained from the base models as it is described above.

Ultimate tensile strength (UTS):

$$\overline{\varepsilon}(t,T) = (b_0 + b_1 e - ^{K_1 t} + b_2 e^{-K_2 t})\overline{\varepsilon}(t,T) + b_3(1 - e^{\overline{\varepsilon}(t,T) b_4}) \tag{5.17}$$

Moduli (MOD), $m = 1,2,..$

$$\overline{\varepsilon}(t,T) = (b_0 + b_1 e - ^{K_1 t} + b_2 e^{-K_2 t}) + b_3(1 - e^{-m b_4}) \tag{5.18}$$

Models presented in Eqs. (5.15-5.18) depend on two common (kinetic) parameters, K_1 and K_2, and several partial (formal) parameters. The common parameters depend on temperature by the Arrhenius law, which can be represented in the following form:

$$K_i = e^{-ki - E_i X}$$

$i = 1,2,$
where

$$X = \frac{1000}{T+273} - X_0, X_0 = \frac{1}{3}\sum_{i=1}^{3}\left(\frac{1000}{T_i + 273}\right) \quad T_1 = 140, T_2 = 125, T_3 = 110$$

This representation has several essential computational advantages (8) in comparison with a conventional form,

$$K = gexp\left(-\frac{F}{R(T+273)}\right) \quad F = 1000RE, g = e^{-k+EX_0} \tag{5.19}$$

This simple transformation helps to improve the model structure and to decrease the influence of multicollinearity, because values E (the activation energies) and k (pre-exponentials) are closer to 1 than their "natural" analogues g and F. (See Table 5.8) As a result, for the AAT data processing we have 4 models with 8 partial and 4 common parameters. To estimate all these parameters, we employ the successive Bayesian estimation (SBE) method (6).

5.9.1 SUCCESSIVE BAYESIAN ESTIMATION

The idea of SBE is very simple. The multi-response regression problem is solved successively when each part of data (series) is analyzed separately, but we have to take into account the information regarding common parameters estimated by the previous series. After processing each data portion, the obtained results are converted into a *posteriori* Bayesian information, which comprises of the vector of parameter estimates and the covariance matrix. As a result, a *posteriori* Bayesian information is formed after each model fitting. Then this information is passed to the next step and used as *a priori* one for fitting the next series. The first series is processed without *a priori* information and the last series provides the ultimate result of the SBE procedure. To form *a priori* information, it is necessary to separate common parameters from the partial ones, and to recalculate the Fisher information matrix, so as to prepare a priori information for the following step. *A priori* information is calculated by modification of the objective function $Q(\mathbf{a})$, which becomes a product of two terms-the sum of squares $S(\mathbf{a})$ and the Bayesian term $B(\mathbf{a})$.

$$Q(\mathbf{a}) = S(\mathbf{a})B(\mathbf{a})$$

The sum of squares S(a) is calculated as follows:

$$S(\mathbf{a}) = \sum_{i=1}^{N} (y_i - f_i)^2 \tag{5.20}$$

and the Bayesian term is given by formula:

$$B(\mathbf{a}) = exp\left(\frac{R(\mathbf{a})}{N}\right) \tag{5.21}$$

In these formulas N is the number of observations. $R(\mathbf{a})$ is calculated as a quadratic form:

$$R(\mathbf{a}) = (\mathbf{a}\text{–}\mathbf{b})^t H(\mathbf{a}\text{–}\mathbf{b})$$

where vector **b** consists of the prior parameter values that are estimates of parameters a obtained at the previous step:

$$\mathbf{b} = (\hat{a}_1, \ldots, \hat{a}_p)^t \tag{5.22}$$

and matrix **H** is the Bayesian information matrix:

$$\mathbf{H} = \left\{h_{ji}, I = 1, \ldots, p, j = 1, \ldots, p\right\} \tag{5.23}$$

which is constructed from the *posteriori* Fisher's matrix **A**:

$$\mathbf{A} = \mathbf{B}(\mathbf{b})\left(\frac{N}{S(\mathbf{b})}\mathbf{V}^t\mathbf{V} + \mathbf{H}\right) \tag{5.24}$$

is calculated in the previous step. Here **V** is the $p \times \mathbf{N}$ matrix, in which elements are the derivatives of the modelling function:

$$V_{ij} = \frac{\partial f(x_j, \mathbf{b})}{\partial a_i} \tag{5.25}$$

$i = 1, \ldots, p, j = 1, \ldots, N$

N is the number of observations. It is important that terms **H** and **b** in Eqs. 5.24 and 5.25 relate to the previous step (series) and they should not be confused with the elements of the same name related to the current step.

When *a priori* information is constructed from a *posteriori* one, it is essential to detach information, which refers to the common parameters, from those that relate to the partial parameters. The common information should be preserved for the next step, but the partial one should be removed since it does not conform with the next portion of data. The *posteriori* Fisher matrix **A** presented in Eq. 5.24 can be written as a block matrix:

$$\mathbf{A} = \begin{bmatrix} A_{00} & A_{01} \\ A_{01} & A_{11} \end{bmatrix}$$

where $\mathbf{A_{00}}$ is the $(q \times q)$ square matrix corresponding to the common parameters, $\mathbf{A_{11}}$ is the $(pq) \times (p–q)$ square matrix corresponding to the partial parameters, and $mathbf{f}_{A_{01}}$ is the $q \times (p–q)$ interrelation matrix. The *a prior* information matrix **H** is recalculated from matrix **A** by a formula:

$$\mathbf{H} = \frac{1}{s^2} \begin{bmatrix} \mathbf{A_{00}} - \mathbf{A_{01}} \mathbf{A_{11}^{-1}} \mathbf{A_{01}^{t}} & 0 \\ 0 & 0 \end{bmatrix} \tag{5.26}$$

where s^2 is the *posteriori* value of the error variance. The dimension of matrix H should correspond to the number of parameters in the next portion of data, so the matrix is completed with the zero values. The *a prior* parameter values are transformed as follows:

$$b_i = \begin{cases} \hat{a} & 0 < i \leq q \\ 0, & q < i \leq p \end{cases} \tag{5.27}$$

Now, Eqs. 5.26, 5.27 present the *a prior* information that is applied to the next step of the SBE procedure.

5.10 RESULTS AND DISCUSSION

The SBE method is employed for the rubber aging example in the following manner. The whole AAT data-set is divided into parts (series) in such a way that every part is related to some response (ELB, DEF). Four common parameters (k_1, E_1, and k_2, E_2) are estimated successively – series by series – using the Bayesian approach. The AAT data are processed in the following order:

$$ELB \xrightarrow{1} DEF \xrightarrow{2} ELB \xrightarrow{3} DEF$$

Table 5.8
Examples of the *posteriori* and the *a priori* information matrices.

Name	Value	Matrix	Matrix	Matrix	Matrix	Matrix	Matrix	Matrix	Matrix	Matrix
b0	1.16e+01	3.01e+03	2.16e+03	2.49e+03	3.41e+02	-1e+05	-2.45e+03	-1.94e+02	1.28e+03	6.24e+01
b1	-4.22e+00	2.16e+03	1.71e+03	1.92e+03	2.41e+02	-8e+04	-1.46e+03	-1.29e+02	6.85e+02	3.62e+01
b2	3.95e+00	2.48e+03	1.92e+03	2.23e+03	2.78e+02	-1e+05	-1.80e+03	-1.81e+02	8.01e+02	5.59e+01
b3	-6.22e+01	3.41e+02	2.41e+02	2.78e+02	3.89e+01	-1e+04	-2.86e+02	-2.23e+01	1.51e+02	7.30e+00
b4	1.40e-01	-1.00e+05	-8.00e+04	-1.00e+05	-1.00e+04	5e+06	1.00e+05	8.00e+03	-6.00e+04	-3.00e+03
k1	1.65e+00	-2.45e+03	-1.46e+03	-1.80e+03	-2.86e+02	1e+05	2.97e+03	2.11e+02	-1.57e+03	-6.85e+01
E1	1.23e+01	-1.94e+02	-1.29e+02	-1.81e+02	-2.23e+01	8e+03	2.11e+02	3.48e+01	-6.85e+01	-8.80e+00
k2	1.51e+00	1.28e+03	6.85e+02	8.01e+02	1.51e+02	-6e+04	-1.57e+03	-6.85e+01	1.13e+03	3.24e+01
E2	2.58e+01	6.24e+01	3.62e+01	5.59e+01	7.30e+00	-3e+03	-6.85e+01	-8.80e+00	3.24e+01	4.20e+00

-	0.00e+00	0e+00	0e+00	0e+00	0.00e+00	0.00e+00	0.00e+00	0.0e+00	-
-	0.00e+00	0e+00	0e+00	0e+00	0.00e+00	0.00e+00	0.00e+00	0.0e+00	-
-	0.00e+00	0e+00	0e+00	0e+00	0.00e+00	0.00e+00	0.00e+00	0.0e+00	-
k1	1.65e+00	0e+00	0e+00	0e+00	3.28e+02	1.98e+01	-7.14e+01	1.6e+00	exclude
E1	1.23e+01	0e+00	0e+00	0e+00	1.98e+01	8.90e+00	4.60e+00	7.0e-01	exclude
k2	1.51e+00	0e+00	0e+00	0e+00	-7.14e+01	4.60e+00	1.78e+02	4.2e+00	
E2	2.58e+01	0e+00	0e+00	0e+00	1.60e+00	7.00e-01	4.20e+00	5.0e-01	

The SBE procedure begins with the ELB modelling. In the first step a simplified model:

$$\overline{\varepsilon}(t,T) = a_0 + a_1 e^{-K_1 t}$$

is employed, because it is impossible to estimate all parameters in the full model given by 5.16 . At the end of the first step, we construct the *posteriori* information (see 5.24), and recalculate it into *a priori* information regarding parameters k_1 and E_1 as shown in 5.26. This *a priori* information is used for the deformation curve fitting. In the second step it becomes possible to estimate four common parameters and pass them in the form of *a priori* information to the next step. Figure 2.3 shows the *posteriori* information matrix obtained after the end of step two, and recalculates *a priori* information matrix ready for application in the third step.

In step three, we return to the ELB model with a goal to specify parameters a_0, a_1, k_2, E_2 and to estimate a_2. In these calculations, parameters k_1 and E_1 are excluded from optimization in term S, which means that we fix:

$$\frac{\partial S}{\partial k_1} \equiv 0 \quad \frac{\partial S}{\partial E_1} \equiv 0$$

In the result of the third step we have all common (kinetic) parameters estimated, and we only have to estimate the values of partial parameters $b_0, \ldots, b_4$. This is performed in the fourth step, where deformation curve is fitted again with all common parameters fixed as the constants.

Table 5.9
Scheme of the successive parameter estimation.

Parameters	ELB1	DEF-1	ELB-2	DEF-2	Values	
a_0	0.97		**0.64**			
a_1	4.19		**4.31**			
a_2			**0.06**			
b_0		11.56		**11.41**		
b_1		-4.22		**-4.14**		
b_2		3.95		**3.87**		
b_3		-62.20		**-59.71**		
b_4		0.14		**0.14**		
k_1	1.40	**1.65**			2.70E12	g_1
E_1	12.53	**12.32**			102.32	F_1
k_2		1.51	**1.50**		1.96E27	g_2
E_2		25.78	**26.16**		217.44	F_2

In the application of this technique, when the same data-set is used several times, it is very important to see to it that at the next estimations, the common parameters are treated as the constants, and that optimization is performed with respect to the partial parameters only. Table 5.8 presents the scheme of the SBE procedure. The ultimate parameters' values are given in bold. It may be seen that the ultimate values of kinetic parameters are obtained at the second step (DEF-1) and these values do not change afterwards. At the third step (ELB-2), we obtain the ultimate values of the partial parameters a_0, a_1, a_2.

The last but one column presents kinetic parameters recalculated in the conventional "physical" units (5.19) as pre-exponentials $g(hr^{-1})$ and the activation energies, $F(KJ)$. Lifespan prediction is developed by means of extrapolation of all rubber properties over temperature T=20°C for the given critical levels. The results are shown in Table 5.10.

Levels of critical values are defined by a tire producer. For each property the mean and the confidence values are determined. The mean value is calculated as the time to achieve the critical level for a developed model with estimated values for model parameters (row 6). The confidence values (row 5) are calculated with respect to the uncertainty in the parameter estimates. They correspond to the 95% one-side confidence intervals for each model– the upper interval for moduli MOD(m), and the lower ones for other characteristics. The lifespan values are determined as the time when the critical level is achieved by a particular model. All these results are also presented in Figure 5.9, plots b)-d) where we show- the mean values (curves 4), the confidence intervals (curves 5), and the corresponding lifespan periods (points 7,8).

In Table 5.10, the initial and the limiting values of all properties (rows 1, 2) are presented. The initial value is calculated as a property value at t = 0:

$$y_0 = y(T = 20^\circ, t = 0)$$

Table 5.10
Prediction of rubber properties at temperature T = 20°C.

Property	ELB	UTS	MOD(1)	MOD(2)	MOD(3)	MOD(4)	MOD(5)
	(-)	(KPa)	(KPa)	(KPa)	(KPa)	(KPa)	(KPa)
Initial value	5.01	25.05	3.10	7.29	12.40	18.33	24.95
Limiting value	0.71	2.04	7.25	15.57	24.83	34.89	45.66
Critical value	3.00	18.00	5.00	10.00	17.00	25.00	32.00
Relative variation	47%	31%	46%	33%	37%	40%	34%
Confidence time (year)	25.00	23.00	20.00	14.00	16.00	18.00	15.00
Mean time (year)	46.00	47.00	45.00	29.00	34.00	38.00	31.00

A limiting value is calculated by formula:

$$y_{lim} = y(T = 20°C, t = 1000yr)$$

The limiting point is defined considering the following issues. ELB goes down monotonically and its limiting value is $\overline{\varepsilon}(\text{inf}) \approx \overline{\varepsilon}(1000yr)$. Moduli λ_m have a maximum, and the UTS, $\sigma(t)$, has an inflection at the same point. Row 4 represents the relative variation for each property, calculated by formula:

$$\left| \frac{y_0 - y_{crit}}{y_{lim} - y_{crit}} \right|$$

It is interesting to compare the lifespan values (rows 5, 6) with the range values given in row 4. It can be seen that the longer lifetimes (ELB and UTS) are related to the more essential changes in the corresponding properties. We suppose that mismatches in the predicted life times are due to wrong critical limits provided by a producer. To verify this assumption we present a plot in which the aging coefficients (AC):

$$AC = \left| \frac{y_t - y_{crit}}{y_{lim} - y_{crit}} \right|$$

are shown in dependence on AAT time. Figure 5.10 shows the corresponding curves (means and confidence) calculated at the practically important time interval. All curves are equal to zero at t = 0, and tend to 1 when t tends to infinity. The ACs for all moduli are overlaid (curve 2). This fact evidently follows from 5.18. A much more interesting observation is that AC for the mean value of ELB is also overlapped with ACs for all mean moduli (curve 2). This fact does not follow from the models given in Eqs. 5.15-5.18, but this looks like an important characteristic of the rubber

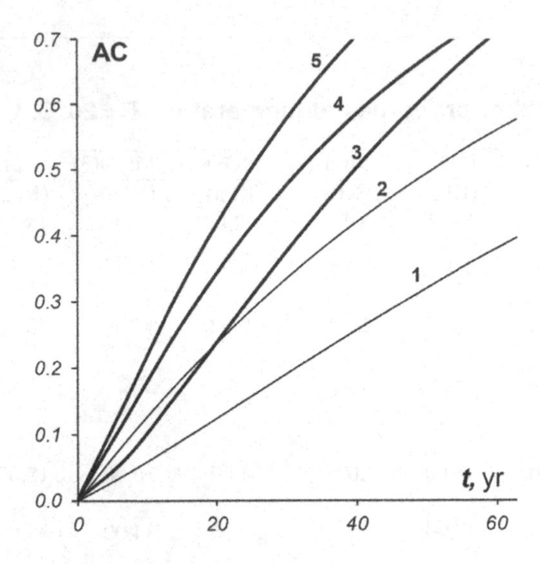

Figure 5.10 Aging coefficients for different properties calculated for the mean and confidence predictions- 1 mean UTS, 2 mean ELB and all mean MODs, 3 confidence UTS, 4 confidence ELB, 5 all confidence MODs.

material. The ACs for UTS (curves 1 and 3) are located lower than other curves. This means that in the course of aging, UTS property changes smaller than any other rubber property.

5.11 CONCLUSIONS

Joint application of EDOE and SBE methods provides evident advantages for prediction of the rubber lifespan. This allows developing a system of models coherently describing various rubber properties. We consider that the existence of such a system, which correctly describes experimental data, proves possibility and validity of the extrapolation.

The crucial point in development of the thermo-oxidation model of aging is the number of common parameters depending on temperature. In our study we employ only two kinetic rate constants, K_1 and K_2, which include four common parameters. It can be demonstrated that rate constant K_2 is required in the DEF model, but is redundant in the ELB model. At the same time, introduction of the third constant K_3 does not improve the UTS modelling, but leads to overfitting. It is known (9) that overfitting results in unreliable prediction. The "redundant" term K_2 in the ELB model does not contradict this principle, as K_2 is estimated using the deformation curve. As a matter of fact, 5.15-5.18 represent a universal multi-response model, where parameters are shared between all the responses.

In particular, the overlap of the AC curves (moduli and ELB in 5.10) is the result of the K_2 redundancy for prediction of these properties. At the same time, it is necessary to include rate K_2 into the ELB model. If this is not done, the estimates of k_1 and E_1 become equal to the values obtained in the first step of the SBE procedure (first column in 5.8). In this case we would obtain overestimated life periods – the mean becomes 52 years, and its confidence value becomes 36 years.

The second issue is the structure of the aging models, and their dependence on the partial parameters. We consider this to be less important than the choice of the common parameters. If we use an appropriate design of AAT and all properties cross their critical levels, we do not need to ever time, and, therefore, the models' shape, i.e., their dependence on time, is not so essential. For example, in the ELB model (5.16) we can use hyperbola instead of exponential, and this replacement does not change the prediction results essentially – the confidence time becomes 25 years instead of 23 years, and the mean time becomes 46 years instead of 43 years.

The most correct way to construct a proper design of experiment is to do it evolutionarily, consequently specify the next measurement point taking into account the previous measurements. We performed this using the EDOE procedure intended for this purpose. It was demonstrated that EDOE is a useful approach that assesses the testing periods for the different aging temperatures and depths before a full-scale AAT is launched.

We can conclude that a proper number of common kinetic parameters can be determined using the principles of adequacy and identifiability. Kinetic parameters have a physical meaning; they reflect the process of thermo-oxidative destruction of polymer. Partial parameters carry no physical meaning and they depend on the selected mathematical model; they are not essential for the prediction results. Analyzing the obtained results of prediction, we can conclude that they are apparently close to the expected lifespan values.

Bibliography

1. E. V. Bystritskaya, A. L. Pomerantsev, and O. Ye Rodionova. Prediction of the aging of polymer materials. In *Chemometrics and Intelligent Laboratory Systems*, volume 47, pages 175–178, may 1999.
2. E. V. Bystritskaya, A. L. Pomerantsev, and O. Ye Rodionova. Evolutionary design of experiment for accelerated aging tests. *Polymer Testing*, 19(2):221–229, 2000.
3. E. V. Bystritskaya, A. L. Pomerantsev, and O. Ye. Rodionova. Non-linear regression analysis: new approach to traditional implementations. *Journal of Chemometrics*, 14(5-6):667–692, sep 2000.
4. E. V. Bystritskaya, A. L. Pomerantsev, and O. Ye. Rodionova. Non-linear regression analysis: new approach to traditional implementations. *Journal of Chemometrics*, 14(5-6):667–692, sep 2000.
5. Alexey L. Pomerantsev. *Chemometrics in Excel*. John Wiley & Sons, Inc., may 2014.
6. Alexey L. Pomerantsev. Successive Bayesian estimation of reaction rate constants from spectral data. *Chemometrics and Intelligent Laboratory Systems*, 66(2):127–139, jun 2003.
7. O. E. Rodionova and A. L. Pomerantsev. On one method of parameter estimation in chemical kinetics using spectra with unknown spectral components. *Kinetics and Catalysis*, 45(4):455–466, jul 2004.
8. O. E. Rodionova and A. L. Pomerantsev. Estimating the parameters of the Arrhenius equation. *Kinetics and Catalysis*, 46(3):305–308, may 2005.
9. Oxana Y. Rodionova and Alexey L. Pomerantsev. Prediction of rubber stability by accelerated aging test modeling. *Journal of Applied Polymer Science*, 95(5):1275–1284, mar 2005.

5.12 PRINCIPAL COMPONENT ANALYSIS APPLIED TO THE STUDY OF THE BEHAVIOR OF STEEL CORROSION INHIBITORS

José Manuel Gandía Romero[3,*] and Pablo Monzón[4]

5.13 INTRODUCTION

Inhibitors are one of the most commonly used methods for reducing the corrosion rate. These are substances that, when added to a corrosive environment, reduce the rate of this process in metals or alloys.

Inhibitors can be applied to all types of corrosion and there are many substances that can act as corrosion inhibitors but, in practice, the number of substances used is very limited (11).

It is common to use inhibitors formed by inorganic compounds such as phosphates. Some of these inhibitors, such as nitrites or chromates, are being replaced by organic compounds due to governmental environmental restrictions. The most widely used organic compounds are alcohols, amines, and polycarboxylates, which, in the same way as mixed inhibitors, are adhered to the metal surface by chemical adsorption and create an organic layer that can restrain both the anodic and cathodic processes. Amines are used because of their high water solubility. Their adsorption to the surface of the metal is produced by their functional group of nitrogen atoms. Polycarboxylate adsorption is influenced by the functional groups present, such as Brønsted-Lowry acids, giving protons (H^+) (2; 3; 20).

Carboxylates and amines are polar organic compounds. They consist of functional groups with localized electric charges. This quality makes them likely to be adsorbed to the metal surface, producing a charge transfer. The size, orientation, shape, and electrical charge of the molecule affect the adsorption extent, and therefore the inhibitor effectiveness (13).

For compatibility reasons, the use of carboxylic acids is common along with other organic bases, usually amines. Thus, the inhibitor is formed with a couple of acid and base compounds. Both behave like a Brønsted-Lowry acid-base, where the amine accepts the proton transferred by the carboxylic acid, and thus the acid and its conjugate base are formed (16). They are considered mixed type inhibitors (20). Mixed inhibitors act on both anodic and cathodic sites and they reduce the corrosion rate without a significant change in the corrosion potential (11; 22).

Cyclic voltammetry is a technique that is commonly used in the study of corrosion processes to determine rupture potentials, metal pitting corrosion behavior, or

[3]Centro de Reconocimiento Molecular y Desarrollo Tecnolgico (IDM), Unidad Mixta Universidad Politcnica de Valencia Universidad de Valencia, Valencia, Spain.

[4]Departamento de Construcciones Arquitectónicas. Universidad Politécnica de Valencia. Camino de Vera s/n. E-46022, Valencia, Spain.

*Corresponding author: joganro@csa.upv.es

the kinetic response of a given system. In this study, the data corresponding to the cyclic voltammograms for each of the studied samples was processed by PCA. The study was performed in aqueous solution with a pH range between 7 and 12.5 in the presence of sulfate anions.

The main objective of this study was to evaluate corrosion inhibitors using traditional techniques and PCA. The study was conducted from neutral to basic pH (12.5) to observe the behavior of the steel in the passivation region.

5.14 MATERIALS AND METHODS

5.14.1 SAMPLES PREPARATION

Corrosion studies were carried out using an inorganic salt Na_2SO_4 at a concentration of $0.1M$ in aqueous solutions. The following organic acids were used as corrosion inhibitors- 4-aminobenzoic acid ($C_7H_7NO_2$), 11-aminoundecanoic acid ($C_{11}H_{23}NO_2$), and 1,10-decanodioic acid (sebacic acid) ($C_{10}H_{18}O_4$). All compounds used were of analytical grade and were purchased from Sigma-Aldrich. Sulfate anions were selected because they produce an increase in the corrosion rate (7; 9; 1).

Salt solutions were prepared at four different pH values- 7, 9, 11, and 12.5. Although the usual pH range in groundwater is between 7 and 9, the studied range was extended up to pH 12.5 in order to benefit from the stability of electrochemical oxides formed at the steel surface. Sodium hydroxide was used to adjust pH solution values.

5.14.2 CHEMICAL SPECIATION EQUILIBRIUM OF INHIBITORS

Table 5.11 shows the bibliographic data of pK_{a1} and pK_{a2} values for the first and second deprotonation of the inhibitors chosen for the study (8; 18).

Table 5.11
Data of pKa values of the three compounds used as inhibitors.

Compound	pKa_1	pKa_2
4-aminobenzoic acid (4AB)	2.50	4.87
sebacic acid (SEBA)	4.72	5.45
11-aminoundecanoic acid (11DECA)	4.55	11.15

Figures 5.11 and 5.12 show the percentages of each of the species in the solution vs pH. The predominant species at each pH establishes the effectiveness of each of the inhibitors (25).

Two of the tested compounds are amino acids. These types of compounds are characterized by an anionic coordinating form with higher pH values, by a poorly coordinating zwitterionic form with intermediate pH values from pK_{a1} to pK_{a2} and by a protonated non-coordinating form with charge $+1$ for $pH < pK_{a1}$. However, for dicarboxylic acid, the pK_a values are very similar to each other, so that at pH greater

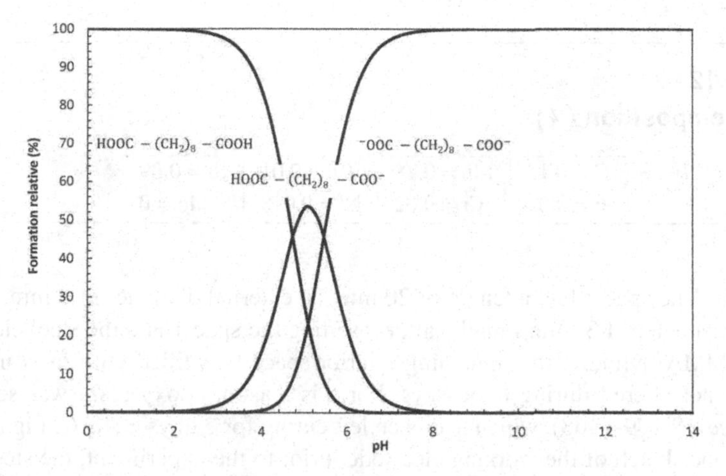

Figure 5.11 Diagram of speciation according to the pH value of sebacic acid.

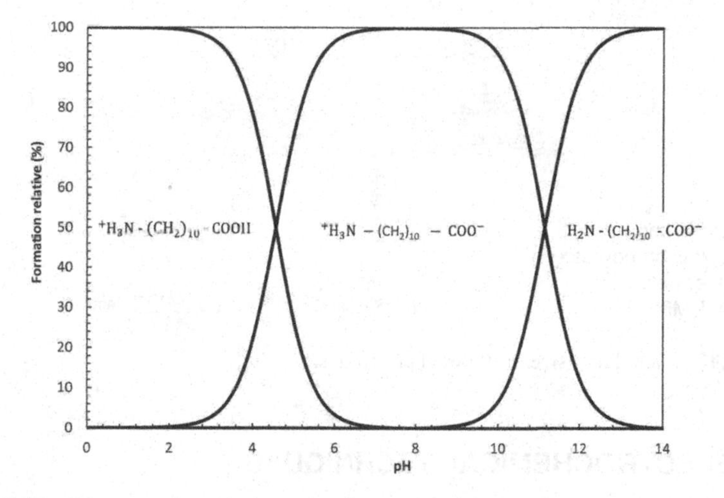

Figure 5.12 Diagram of speciation according to the pH value of 11-aminoundecanioic acid.

than 5.45, coordinating dianion is the predominant form, which should create great stability against corrosion on the metallic surface.

5.15 ELECTRODE PREPARATION

Working electrodes (WE) were made from S355JR steel, which is normally used as structural steel. The composition of the steel is given in 5.12. The final diameter of the electrodes was 4.7 mm with a cross-section of 17.3 mm^2.

The four electrodes were encapsulated in Poly(methyl)Methacrylate (PMMA) cylinders with internal diameter of 8 mm, external diameter of 10 mm, and a length

Table 5.12
Steel Composition (%).

C = 0.13	Mn = 0.78	Cu = 0.01	Si = 0.39
P = 0.02	Cr = 0.02	Ni = 0.018	Mo = 0

of 64 mm. The spacer had a length of 20 mm, an external diameter of 8 mm, and an internal diameter of 3 mm. Finally, after inserting the spacer and the steel electrode in the PMMA cylinder, the remaining interior space was filled with inert material that was not altered during the assays. For this reason, epoxy resin was selected, (reference RS 199-1468), which had been left curing for 2 days at 40°C. Figure 5.13 displays the sketch of the working electrode. Prior to the experiment, the steel electrodes were cleaned with distilled water, abraded with abrasive paper, and decreased in methanol.

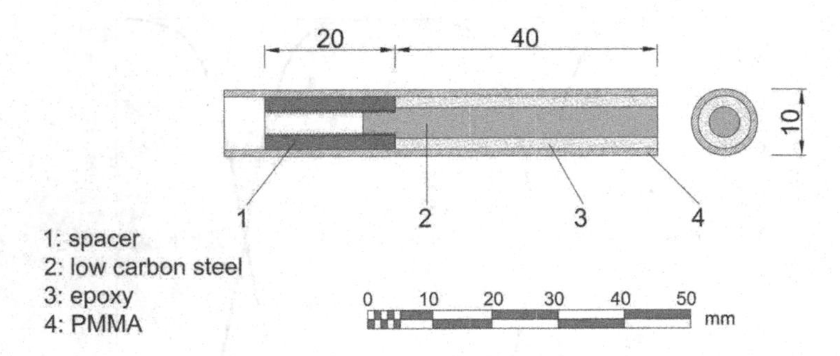

Figure 5.13 Detail and size of the steel electrodes.

5.16 ELECTROCHEMICAL TECHNIQUES

Electrochemical experiments were performed using an Autolab PGSTAT 100 instrument (from EcoChemie, NL). Aqueous solutions of Na_2SO_4 $0.1M$ were studied at four different pH values: 7, 9, 11, and 12.5.

The three studied inhibitors were added to the aqueous solution in a concentration of 7% w/w. Each prepared solution was measured four times, with all four different homemade electrodes, giving a total of 64 samples (1 salt solution $\times$ (3 inhibitors + 1 without inhibitor) $\times$ 4 pH values $\times$ 4 electrodes).

All measurements were performed under synthetic air (79% N2, 21% O2) at a temperature of 25.0 ± 0.1°C using a PolyScience 9106 temperature controller. The electrochemical experiments were carried out in a conventional three-electrode cell. A platinum wire electrode and a Saturated Calomel Electrode (SCE) were used as counter and reference electrode, respectively.

5.16.1 ZERO CURRENT POTENTIAL AND POTENTIODYNAMIC POLARISATION MEASUREMENT

The half-cell potential of the electrode from each sample was monitored versus time. Corrosion potential (Ecorr) was obtained after a stabilization time (it was usually achieved in the range of 8 to 10 hours).

When the half-cell potentials recorded from the different electrodes were found to be stable (it was considered that the E_{corr} had been achieved when the potential drift was less than $0.1\mu V\ s^{-1}$), the technique of potentiodynamic polarisation was applied. Potential scans began 140 mV below the E_{corr}/OCP up to a value of 140 mV at a scan rate of 0.5 mV/s. The linear Tafel segments were extrapolated to the point of intersection to obtain the corrosion potential (E_{corr}) and corrosion currents density (i_{corr}) (19; 4).

5.17 CYCLIC VOLTAMMETRY

After the potentiodynamic polarization measurement, the electrode's surface was prepared. Then cyclic voltammetry tests were performed, and the steel stabilisation period was controlled in a thermostatized cell under synthetic air. Experiments were carried out with static electrodes. For these voltammetric studies, two consecutive scans were carried out. For the statistical analysis, only the data corresponding to the second scan was used because of their higher reproducibility. Potential scans were from -0.9 V to +0.5 V at a scan rate of 10 mVs^{-1}.

5.18 DATA MANAGEMENT MULTIVARIATE ANALYSIS

A data matrix was created where *the number of objects was decided by the number of experiments and the variables were the current responses created by the applied potential* (21; 14; 23).

In this statistical analysis data preprocessing was necessary. Various preprocessing methodologies were used to choose the most adequate method. In the end, the *autoscale method* was applied. This method combines centering and standardization transformations. In this case, *it was possible to give importance* to the intensity measurements of samples with passivation and produce a better separation of the clusters. After this preprocessing method, the data matrix was standardized since standardization gives all the variables the same variance, and all the variables had the same influence on the estimation of the components.

All statistical analysis was performed with the software SOLO (version 6.5, Eigenvector Research, Inc.).

5.19 RESULTS AND DISCUSSION

5.19.1 OPEN CIRCUIT POTENTIAL (OCP) AND TAFEL POLARIZATION MEASUREMENT

Tafel extrapolation method allows the analysis of the current versus potential curves. This analysis provides information about the system's features by means of two basic data- the equilibrium potential value and the current value of the electrochemical corrosion. Various parameters, such as corrosion potential (E_{corr}), corrosion current density (i_{corr}), anodic Tafel slope (βa), cathodic Tafel slope (βc), and inhibition efficiency are summarized in 5.13.

The inhibition efficiency is calculated from the value of corrosion current density, which can be evaluated from extrapolation method, using the following equation (17; 6; 12):

$$\eta = (i^o_{corr} - i_{corr})/(i^o_{corr})x100$$

where i^o_{corr} and i_{corr} are the corrosion current densities in the absence and presence of inhibitors.

Table 5.13
Polarization parameters for steel electrodes.

Treatment	pH	Ecorr vs SCE (V)	βa (V/dec)	βc (V/dec)	icorr (μA/cm2)	vi (%)
Without inhibitor	7	-0.707	0.120	-0.706	29.10	-
	9	-0.675	0.097	-0.927	25.04	-
	11	-0.678	0.130	-0.559	20.67	-
	12.5	-0.177	0.184	-0.240	0.080	-
4AB + SO_4^{-2}	7	-0.672	0.108	-0.462	20.31	30.2
	9	-0.668	0.116	-0.597	17.66	29.5
	11	-0.682	0.100	-0.485	13.33	35.5
	12.5	-0.168	0.246	-0.193	0.079	1.3
SEBA + SO_4^{-2}	7	-0.694	0.125	-0.392	19.23	33.9
	9	-0.691	0.113	-0.370	15.82	36.8
	11	-0.685	0.098	-0.263	12.07	41.6
	12.5	-0.202	0.287	-0.161	0.075	6.1
11DECA + SO_4^{-2}	7	-0.668	0.095	-1.025	26.63	8.5
	9	-0.664	0.113	-0.619	21.95	12.3
	11	-0.675	0.085	-0.890	18.96	8.3
	12.5	-0.185	0.267	-0.177	0.070	12.5

Figure 5.14 shows the variation in the average corrosion potential (E_{corr}) of the steel electrodes versus the pH for Na_2SO_4 $0.1M$. The three remaining lines show the average OCP value of electrodes in the presence of the different chemical inhibitors studied. Figure 5.14 shows that the addition of chemical inhibitors does not produce significant E_{corr} variations. This behavior is normal for mixed type inhibitors (11; 22).

Figure 5.15 also shows the average corrosion intensity of four electrodes studied in the presence 0.1 M of sodium sulfate at different pH values. In the absence of a chemical inhibitor, corrosion intensity had a maximum value of 29.10 $\mu A/cm^2$ at pH 7, which decreased with increasing pH (25.04 and 20.67 μA/cm2 at pH 9 and 11, respectively) up to a value of $0.08\mu A/cm^2$ at pH 12.5 (5; 15; 24)).

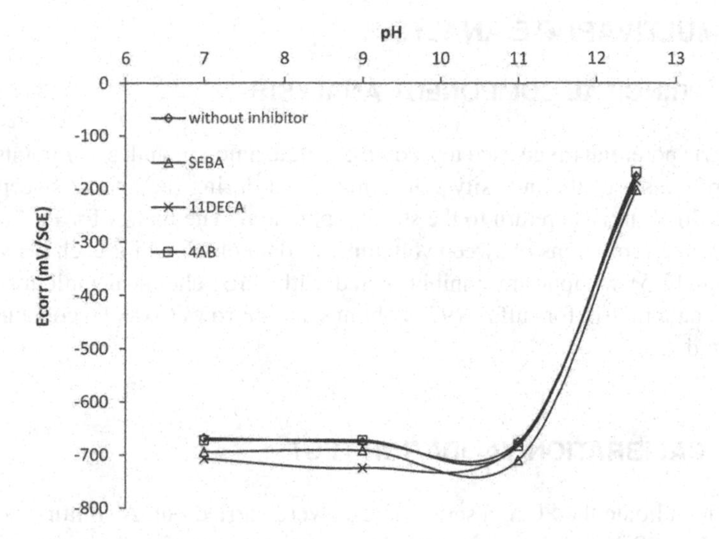

Figure 5.14 Average corrosion potential (E_{corr}) value of the steel electrodes in $Na2SO4$ $0.1M$ with respect to varying pH in the absence and presence of 7 w/w of inhibitor.

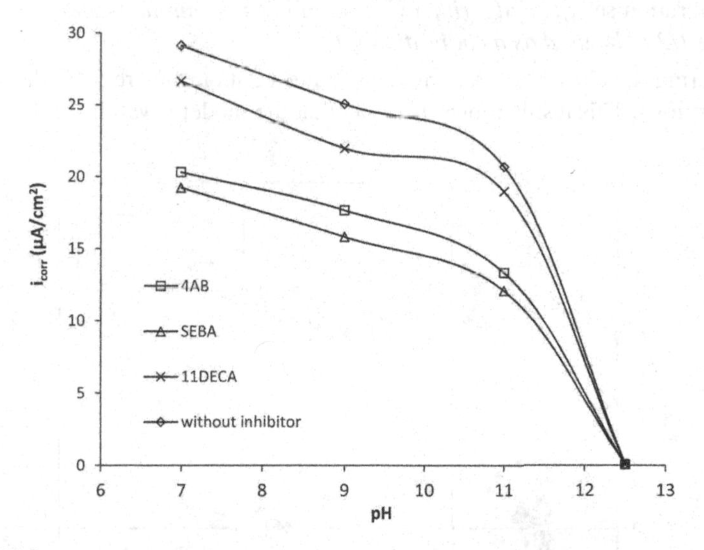

Figure 5.15 Average corrosion intensity of the steel electrodes in Na_2SO_4 $0.1M$ with respect to varying pH in the absence and presence of 7 w/w of inhibitor.

Figure 5.15 shows that the addition of sebacic acid and 4-aminobenzoic acid produced a significant decrease in the corrosion rate for all media studied. On the other hand, 11-aminoundecanoic acid behaved worse when pH values decreased (at these values the predominant species is not the most appropriate for good performance against corrosion).

5.20 MULTIVARIATE ANALYSIS

5.20.1 PRINCIPAL COMPONENT ANALYSIS

The electric potential range used to study the sulfate anion includes 977 points. These points are the set of the intensity values measured during the anodic sweep from -0.8 to +0.36 V until its return to the starting potential. The matrix for PCA analysis includes three repetitions of sweep voltammetry data obtained for each pH value (7, 9, 11, and 12.5) without any inhibitors and with three chemical inhibitors tested. Thus, the data matrix for sulfate (977 columns and 48 rows (3x4x4)) contains a total of 46,896 data.

5.20.2 CALIBRATION-VALIDATION TEST

In order to validate the PCA, a series of tests were carried out. A splitting algorithm integrated into SOLO software was used to separate the whole data set into a calibration set and a validation set. The software permits us to choose between two splitting algorithms. The one used is called the onion method (it keeps out covariance samples plus random inner-space samples and the user has to decide the percentage to keep in the calibration set). *The algorithm was set to keep 2/3 in the training set and the remaining 1/3 to be used as a calibration set.*

Comparing the clusters of training data and test data in Figure 5.16, the similarities are obvious. This result allows us to say that the model is valid.

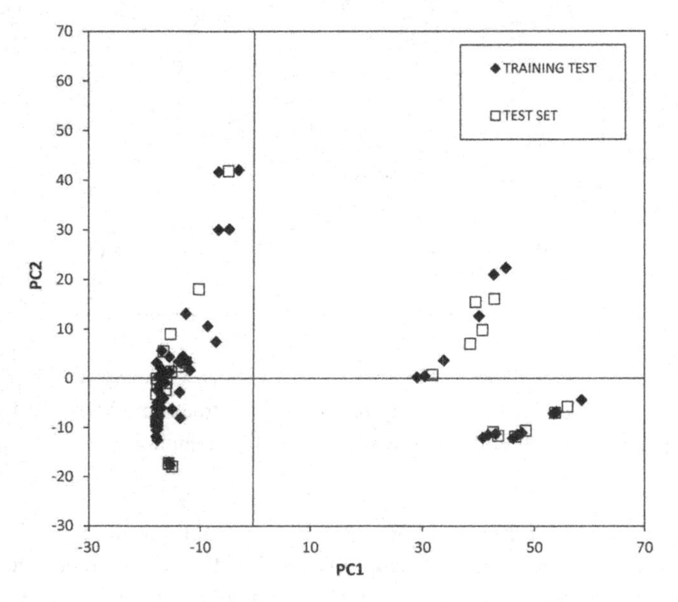

Figure 5.16 Calibration and validation data-sets.

5.20.3 CYCLIC VOLTAMMETRY STUDY

Figures 5.17 and 5.18 show the loading plots of the three principal components. In these plots, the Y-axis shows the loads of the variables of each principal component (PC1, PC2, and PC3). The X-axis shows the variables used in the study. In this case, 977 values of electric potential were seen ($-0.8V$ /$+0.36V$/$-0.8V$), which was the potential sweep range used for this study. With only these three components (PC1, PC2, and PC3), 96.8% of the total variance observed in the system is explained.

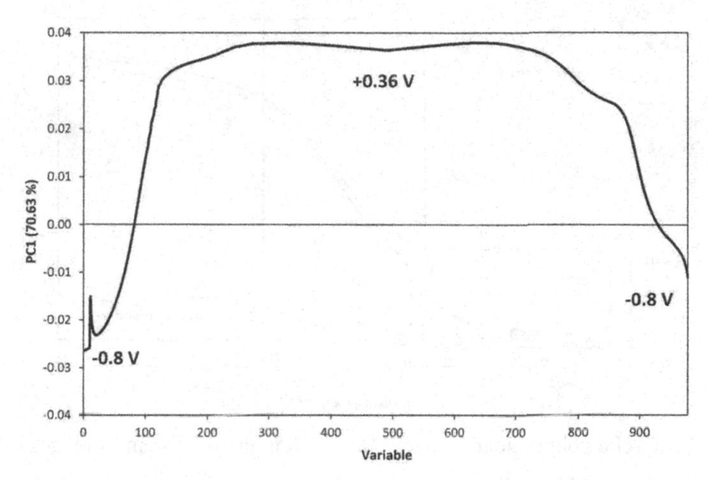

Figure 5.17 Loading plots for PC1 for steel in the presence of different chemical inhibitors and $Na_2SO40.1M$.

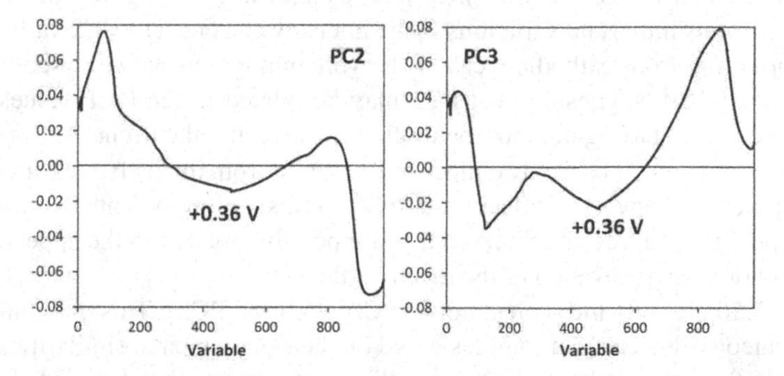

Figure 5.18 Loading plots for PC2 and PC3 for steel in the presence of different chemical inhibitors and 0.1 M $Na2SO4$.

The graphical representation of statistical loads of PC1 (70.63%) indicates that the information which contributes most in this component is obtained from the current intensities obtained for variables between 150 and 850-350 values corresponding to an anodic potential (between $-0.40/+0.36V$) and other 350 values for the cathode sweep ($+0.36/-0.4V$). Statistical loads of PC2 (16.64%) show that the most significant information is extracted from two potential ranges located on the ends of the sweep: -0.8 to -0.4 (anodic) and -0.4 to $-0.8V$ (cathodic). The main component PC3 (9.53%) extracts information mainly from the range of potential within currents between -0.15 and -0.62 V (anodic and cathodic sweep).

Figure 5.19 shows the correspondence between the voltammetric intensity curves against the potential and information on values related to PCA components.

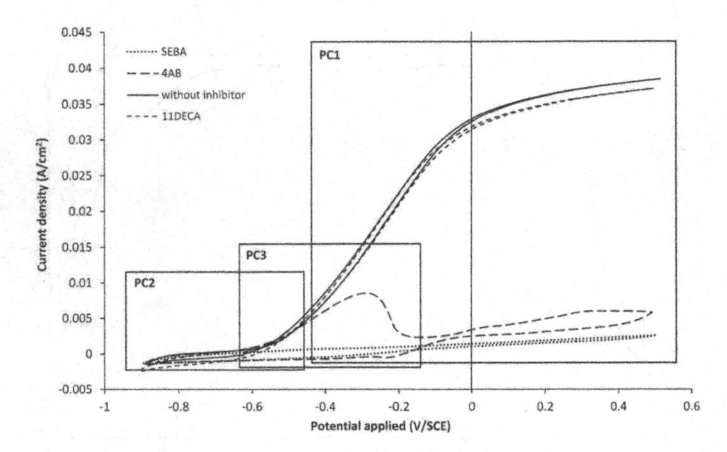

Figure 5.19 Cyclic voltammograms for sulfate/iron/inhibitor system of ranges of the load plots.

This figure suggests that the main component of PC1 is related to the magnitude of the intensity of corrosion ($I_c orr$). In fact, as shown in the figure, this range of potential presents important variations in the intensity current. The PC2 variable collects information from cathodic areas of the voltammograms as we observed in an earlier study ((10)), suggesting that PC2 may be related to the OCP values or the curve shape at the starting point of the discharge of electrical current.

The PC3 component basically collects information from intensity measured in the range of potential between -200 and -600 mV. In this region, potential forms of the intensity/potential curves are observed that are possibly related to the appearance of passivation or transpassivation of the metal surface.

Figure 5.20 presents the scores plot of PCA (PC1 vs. PC2). This diagram shows the spontaneous clustering of samples based on their fundamental similarities or differences. In this case we have plotted six ellipses to separate the observed clusters.

The validity of the marked groupings was tested using an analysis algorithm (called the onion method), which is integrated into the software SOLO for determining reliability.

Figure 5.20 shows that these groups (C, D, E, B2), with lower corrosion rates, are on the left side of the diagram with PC1 values < 0. The separation of the four clusters is facilitated by the values of the second principal component PC2, which are very different among themselves (the scores vary between -20 to +40).

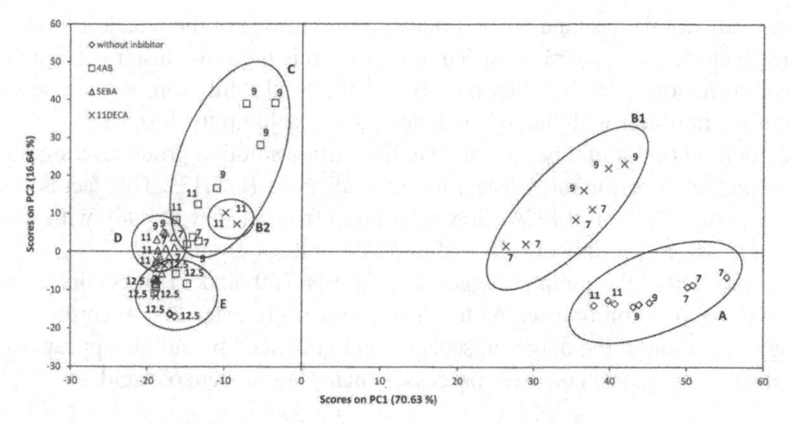

Figure 5.20 PCA of voltammetric data of steel in 0.1 M of Na_2SO4 in the presence of the chemical inhibitors studied (the numbers indicate the pH values).

The other two clusters A and B1 are associated with high corrosion rates (see 5.13). Groupings appear on the right side of the diagram (PC1>0), and also have great differences with respect to the PC2. If we look at Figure 5, it confirms that, as we have discussed above, PC1 is related to the corrosion rate, while PC2 seems to be related to potential where the electric transfer starts apprcciably (a type of breakdown potential).

The data associated with steel corrosion when it is only in the presence of 0.1 M of sulfate is located in the right lower quadrant of the diagram of PCA (Cluster A). The cluster points move to the left region (the corrosion rate decreases) when pH increases. In this cluster (A), the maximum value observed for steel corrosion rate is 29.10 $\mu A/cm^2$ at pH 7. For pH 9, the rate decreases to 25.04 $\mu A/cm^2$ and at pH 11 to 20.64 $\mu A/cm^2$. This is in agreement with the discussion conducted above concerning the loading plots where the principal component (PC1) is related to the corrosion rate.

When the pH of the solution reaches the value of 12.5, it inhibits the corrosion of metal, and there is a balance of points obtained from cluster A to cluster E.

Cluster E, located in the left lower quadrant of the diagram, is a region of low corrosion rate or metal passivation, with corrosion intensities less than 0.08 $\mu A/cm^2$ (5.13). This situation has been discussed in an earlier study (10) and is mainly imposed by the pH (12.5) of the solution.

When 11-aminoundecanoic acid is added in a concentration five times less than the sulfate anion, a typical electrochemical behavior that shows the weak capacity of the compound for inhibiting the corrosion processes of steel can be observed. The anomalous distribution of points (Clusters B1, B2, and E) is corroborated by measuring the corrosion intensity obtained for the same pH using the Tafel extrapolation method.

This behavior occurs because this inhibitor is in the zwitterionic form, which is weakly coordinaing and is the dominant form (more than 90%) in the pH range between 4.5 and 10.5 (see Figure 2). Starting at pH 11.15 (pKa2), the coordinating shape already dominates, and when it binds to the surface of the metal, it produces a jump from cluster B1 (where the speed of corrosion is high) to cluster B2 (where the speed of corrosion is low). When the pH is 12.5, total inhibition occurs due to the action of the inhibitor with the pH, and there is a new jump to cluster E.

The addition of 4-aminobenzoic acid to the sulfate solution produces a significant decrease in the corrosion rate of steel for any value of pH (5.13). This fact is clearly reflected in the diagram of PCA showing a jump from cluster A (only with sulfate) to cluster C when 4-aminobenzoic acid is added to the system.

For sebacic acid the jump is bigger, and for pH 7, 9, and 11, it occurs from the initial starting position (cluster A) to a new position (cluster D). According to the topology evolution in the diagram, sebacic acid enhanced by sulfate appears to stabilize steel more against corrosive processes than 4-aminobenzoic acid.

5.21 CONCLUSIONS

The effectiveness of 4-aminobenzoic acid, 11-aminoundecanoic acid, and sebacic acid as inhibitors of corrosion in steel specimens immersed in solutions with sulfate ions was studied at different pH values using classical electrochemical techniques (OCP, potentiodynamic polarization, and cyclic voltammetry).

The PCA method was applied to analyze the data obtained through sweep cyclic voltammetry. This method allows us to study the influence of the selected chemical inhibitors on the morphology of the voltammograms of rebar in the absence and presence of selected chemical inhibitors. The statistical analysis of PCA generates clusters of points that depend on the pH and the inhibitor admixture used.

The study indicates that the performance of 4-aminobenzoic acid and sebacic acid is better against corrosion than 11-aminoundecanoic acid. In the ranges of pH used in the study, the former two acids are always in the most appropriate species. On the other hand, 11-aminoundecanoic acid does not have a desired behavior in two more acidic pH values (7 and 9) because the predominant species is not adequate (a poorly coordinating zwitterionic form).

Using the data registered in the cyclic voltammograms jointly with the statistical technique of PCA, it is possible to provide more information about the behavior of each one of the studied inhibitors, such as phases of passivation or behavior of the predominant species of the inhibitor depending on the pH. Analyzing all the data obtained in a cyclic voltammogram provides more information about the OCP zone (middle) or extreme ranges. The technique of PCA allows this type of analysis.

Inhibitors that have similar behavior when they are analyzed with the technique of potentiodynamic polarization, may behave differently after analyzing the current and potential values with PCA, which is why it can be concluded that the combination of both techniques provides additional information.

Bibliography

1. E. E. Abd El Aal, S. Abd El Wanees, A. Diab, and S. M. Abd El Haleem. Environmental factors affecting the corrosion behavior of reinforcing steel III. Measurement of pitting corrosion currents of steel in Ca(OH)2 solutions under natural corrosion conditions. *Corrosion Science*, 2009.
2. A. S. Abdulrahman, Mohammad Ismail, and Mohammad Sakhawat Hussain. Corrosion inhibitors for steel reinforcement in concrete: A review, sep 2011.
3. Ahmed M. Al-Sabagh, Notaila M. Nasser, Ahmed A. Farag, Mohamed A. Migahed, Abdelmonem M.F. Eissa, and Tahany Mahmoud. Structure effect of some amine derivatives on corrosion inhibition efficiency for carbon steel in acidic media using electrochemical and Quantum Theory Methods. *Egyptian Journal of Petroleum*, 2013.
4. Mohammed A. Amin, K. F. Khaled, and Sahar A. Fadl-Allah. Testing validity of the Tafel extrapolation method for monitoring corrosion of cold rolled steel in HCl solutions - Experimental and theoretical studies. *Corrosion Science*, 2010.
5. ASTM. ASTM G102 - 89(2010) Standard Practice for Calculation of Corrosion Rates and Related Information from Electrochemical Measurements. In *Annual Book of ASTM Standards*. 2010.
6. Zhenyu Chen, Lingjie Li, Guoan Zhang, Yubing Qiu, and Xingpeng Guo. Inhibition effect of propargyl alcohol on the stress corrosion cracking of super 13Cr steel in a completion fluid. *Corrosion Science*, 69:205–210, apr 2013.
7. Liberato Ciavatta, Gaetano De Tommaso, and Mauro Iuliano. Stability constants of iron(II) sulfate complexes. *Annali di chimica*, 92(5-6):513–20.
8. Marvin L. Deviney, R. C. Anderson, and W. A. Felsing. Application of the Glass Electrode to the Determination of the Thermodynamic Ionization Constants of p-Aminobenzoic Acid. *Journal of the American Chemical Society*, 79(10):2371–2373, may 1957.
9. M. Echeverría Boan, A. Rodríguez, C. M. Abreu, and C. A. Echeverría. Unraveling the impact of chloride and sulfate ions collection on atmospheric corrosion of steel. *Corrosion*, 69(12):1217–1224, dec 2013.
10. J. M. Gandía, P. Monzón, R. Bataller, I. Campos, J. Manuel Lloris, and J. Soto. Principal component analysis applied to study of carbon steel electrochemical corrosion. *Corrosion Engineering Science and Technology*, 50(4):320–329, may 2015.
11. L. A. Godínez, Y. Meas, R. Ortega-Borges, and A. Corona. Los inhibidores de corrosión. *Revista de Metalurgia (Madrid)*, 39(2):140–158, mar 2003.
12. Xinkuai He, Yumei Jiang, Chen Li, Weichun Wang, Bailong Hou, and Luye Wu. Inhibition properties and adsorption behavior of imidazole and 2-phenyl-2-imidazoline on AA5052 in 1.0M HCl solution. *Corrosion Science*, 83:124–136, 2014.
13. M. G. Hosseini, M. Ehteshamzadeh, and T. Shahrabi. Protection of mild steel corrosion with Schiff bases in 0.5 M H2SO4 solution. *Electrochimica Acta*, 52(11):3680–3685, mar 2007.

14. Patrik Ivarsson, Susanne Holmin, Nils Erik Höjer, Christina Krantz-Rülcker, and Fredrik Winquist. Discrimination of tea by means of a voltammetric electronic tongue and different applied waveforms. In *Sensors and Actuators, B: Chemical*, volume 76, pages 449–454, jun 2001.
15. Joint Committee for Guides in Metrology. Evaluation of measurement data — Supplement 1 to the "Guide to the expression of uncertainty in measurement" — Propagation of distributions using a Monte Carlo method. *Evaluation*, 2008.
16. P. Kern and D. Landolt. Adsorption of organic corrosion inhibitors on iron in the active and passive state. A replacement reaction between inhibitor and water studied with the rotating quartz crystal microbalance. *Electrochimica Acta*, 47(4):589–598, nov 2001.
17. Florian Mansfeld. Tafel Slopes and Corrosion Rates from Polarization Resistance Measurements. *Corrosion*, 29(10):397–402, oct 1973.
18. Arthur E. Martell and Robert M. Smith. *Critical Stability Constants*. Springer US, 1982.
19. E. McCafferty. Validation of corrosion rates measured by the Tafel extrapolation method. *Corrosion Science*, 47(12):3202–3215, dec 2005.
20. Marco Ormellese, Luciano Lazzari, Sara Goidanich, Gabriele Fumagalli, and Andrea Brenna. A study of organic substances as inhibitors for chloride-induced corrosion in concrete. *Corrosion Science*, 51(12):2959–2968, dec 2009.
21. Matteo Scampicchio, Simona Benedetti, Barbara Brunetti, and Saverio Mannino. Amperometric electronic tongue for the evaluation of the tea astringency. *Electroanalysis*, 18(17):1643–1648, sep 2006.
22. T. A. Söylev and M. G. Richardson. Corrosion inhibitors for steel in concrete: State-of-the-art report. *Construction and Building Materials*, 22(4):609–622, apr 2008.
23. Kiyoshi Toko. Taste sensor. *Sensors and Actuators, B: Chemical*, 64(1-3):205–215, jun 2000.
24. Laboratory measurement of corrosion speed using the polarization resistance technique. Standard, UNE, Asociación Española de Normalización,, Madrid, ES, 2011.
25. B. Van de Graaf, A. J. Hoefnagel, and B. M. Wepster. Substituent effects. 7. Microscopic dissociation constants of 4-amino- and 4-(dimethylamino)benzoic acid. *The Journal of Organic Chemistry*, 46(4):653–657, feb 1981.

A Software Workflow

A.1 SOFTWARE

As reported in the book "Implementing Reproducible Research" by Victoria Stodden et al. (CRC Press, Taylor and Francis, 2014), reproducibility is the ultimate standard by which scientific findings are judged.

Several solutions have now been implemented in order to let scientists write and share code in the form of interactive *notebooks*, where the calculations are performed transparently in front of the reader, who can also change it. Also, these notebooks do not confine the writer to one software language but offer a plethora of solutions. Python and R, widely adopted by scientist, are among these.

This book uses the R language.

R, as described in the R project website: "is a language and environment for statistical computing and graphics. It is a GNU project which is similar to the S language and environment which was developed at Bell Laboratories (formerly ATT, now Lucent Technologies) by John Chambers and colleagues. R can be considered as a different implementation of S. There are some important differences, but much code written for S runs unaltered under R.

R provides a wide variety of statistical (linear and nonlinear modelling, classical statistical tests, time-series analysis, classification, clustering) and graphical techniques, and is highly extensible. The S language is often the vehicle of choice for research in statistical methodology, and R provides an Open Source route to participation in that activity".

Thanks to its extensibility, several specialized libraries were used. Here are more details:

tidyverse in order to clean and prepare the data-set, unfold matrices to prepare plots. In the words of its creator: "The tidyverse encompasses the repeated tasks at the heart of every data science project: data import, tidying, manipulation, visualization, and programming" (10).

caret package for performing regression on the data. From the package description, "The caret package, short for classification and regression training, contains numerous tools for developing predictive models using the rich set of models available in R. The package focuses on simplifying model training and tuning across a wide variety of modeling techniques. It also includes methods for pre-processing training data, calculating variable importance, and model visualizations" (4).

FactoMineR to perform multivariate analysis. From the description of the package: "FactoMineR is an R package dedicated to multivariate data analysis. The main features of this package is the possibility to take into account different types of variables (quantitative or categorical), different types of structure on the data (a partition on the variables, a hierarchy on the variables, a partition on the individuals), and finally supplementary information (supplementary individuals and variables). Moreover, the dimensions issued from the different exploratory data analyses can be automatically described by quantitative and/or categorical variables. Numerous graphics are also available with various options" (26).

rsm to perform response surface modelling. It is "designed to provide R support for standard response-surface methods. Functions are provided to generate central-composite and Box-Behnken designs. For analysis of the resulting data, the package provides for estimating the response surface, testing its lack of fit, displaying an ensemble of contour plots of the fitted surface, and doing follow-up analyses, such as steepest ascent, canonical analysis, and ridge analysis" (6).

multcomp for performing Principal Component analysis. It "extends the canonical theory of multiple comparison procedures in ANOVA models to linear regression problems, generalized linear models, linear mixed effects models, the Cox model, robust linear models, etc. Several examples using a variety of different statistical models illustrate the breadth of the results" (3).

FrF2 for the DOE chapter. FrF2 is a package "for design and analysis of experiments with 2-level factors. The package offers both regular and non-regular fractional factorial 2- level designs, in the regular case with blocking and split plot facilities and algorithms for ensuring estimability of certain two-factor interactions" (1).

mixexp "provides functions for creating mixture designs composed of extreme vertices and edge and face centroids in constrained mixture regions where components are subject to upper, lower, and linear constraints" (5).

knitr is used to prepare the notebooks for performing calculations in the book, and also to ensure reproducibility of the results (11).

The plots were prepared using the native plot functions present in R, and also with the help of **ggplot2**, **lattice** (7), and**plot3D** (8). For preparing the tables **kable** and **stargazer** (2) were used.

Even if in the editing of this book the use of **RStudio** (9), and packages included in tidyverse (and in specific ggplot2 which is included in the tidyverse) was intensive, it was preferred to report workflows that make a parsimonious use of functions not included in the base distribution of R and that make use of its base plotting capabilities whenever possible.

The following script was tested on a windows system on R version 3.6.3 (2020-02-29). Please note that they are not written by a computer scientist, but by a chemist, and so this code will probably look cumbersome or not "clean", even if it indeed works.

Bibliography

1. Ulrike Grömping. R package FrF2 for creating and analyzing fractional factorial 2-level designs. *Journal of Statistical Software*, 2014.
2. Marek Hlavac. Ready-Made Regression Tables from the Stargazer Package in Statistics Education. *SSRN Electronic Journal*, 2014.
3. Torsten Hothorn, Frank Bretz, and Peter Westfall. Simultaneous inference in general parametric models, 2008.
4. Max Kuhn. caret Package. *Journal Of Statistical Software*, 2008.
5. John Lawson and Cameron Willden. Mixture experiments in R using mixexp. *Journal of Statistical Software*, 2016.
6. Russell V. Lenth. Response-surface methods in R, using RSM. *Journal of Statistical Software*, 2009.
7. Deepayan Sarkar. Package 'lattice': Trellis Graphics for R. *R documentation*, 2017.
8. Karline Soetaert. plot3D, 2019.
9. RStudio Team. RStudio: Integrated Development for R studio Inc, 2005.
10. Hadley Wickham, Mara Averick, Jennifer Bryan, Winston Chang, Lucy McGowan, Romain François, Garrett Grolemund, Alex Hayes, Lionel Henry, Jim Hester, Max Kuhn, Thomas Pedersen, Evan Miller, Stephan Bache, Kirill Müller, Jeroen Ooms, David Robinson, Dana Seidel, Vitalie Spinu, Kohske Takahashi, Davis Vaughan, Claus Wilke, Kara Woo, and Hiroaki Yutani. Welcome to the Tidyverse. *Journal of Open Source Software*, 2019.
11. Yihui Xie. Knitr: A comprehensive tool for reproducible research in R. In *Implementing Reproducible Research*. 2014.

In order to run the script, the following libraries MUST be loaded. Libraries required for loading:

```
library(knitr)
library(qqplotr)
library(reshape2)
library(tidyr)
library(ggplot)
library(ggpubr)
library(multcomp)
library(FactoMineR)
library(factoextra)
library(zoo)
library(GGally)
library(DOE)
library(rsm)
```

A.2 CHAPTER 1

To load the data-set

```
load(file = "myDrive/MyPath/Epoxy.RData")
```

In order to see the structure of the data

```
str(all_data)
```

```
## 'data.frame':    36 obs. of  4 variables:
##  $ A: num  32.2 38.1 41.3 27.6 38.7 ...
##  $ B: num  46.2 34.6 40.2 41.8 40.4 ...
##  $ C: num  38.9 33.6 39.1 33.6 45.8 ...
##  $ D: num  36.9 32.7 41.2 39.1 36.9 ...
```

We can see that the data is a dataframe of four sets A, B, C, D, which represents the observation made by four laboratories on the same variable. Due to the kind of object we are using in R (dataframe), we get the info that we have 36 observations in 4 variables. We calculate a statistical summary of our data

```
summary(all_data)
```

```
##        A               B               C
##  Min.   :27.62   Min.   :22.06   Min.   :23.82
##  1st Qu.:33.42   1st Qu.:35.22   1st Qu.:33.39
```

```
##  Median :34.98   Median :37.53   Median :37.29
##  Mean   :35.64   Mean   :38.41   Mean   :36.75
##  3rd Qu.:37.68   3rd Qu.:42.72   3rd Qu.:41.31
##  Max.   :46.66   Max.   :51.72   Max.   :49.87
##       D
##  Min.   :29.55
##  1st Qu.:33.51
##  Median :37.25
##  Mean   :37.07
##  3rd Qu.:40.45
##  Max.   :47.53
```

In order to calculate the variance

```
var(all_data$A)
```

```
## [1] 12.94204
```

```
var(all_data$B)
```

```
## [1] 52.46142
```

```
var(all_data$C)
```

```
## [1] 40.02367
```

```
var(all_data$D)
```

```
## [1] 17.78311
```

For a stem and lef plot

```
stem(all_data$A, scale = 1)
```

```
##
##   The decimal point is at the |
##
##   26 | 6
##   28 |
##   30 | 52
##   32 | 239034479
##   34 | 22778000112
##   36 | 6935
```

```
##   38 | 17803
##   40 | 834
##   42 |
##   44 |
##   46 | 7
```

```
stem(all_data$B, scale = 1)
```

```
##
##   The decimal point is 1 digit(s) to the right of the |
##
##   2 | 23
##   2 | 778
##   3 | 044
##   3 | 55666677778
##   4 | 001112224
##   4 | 566679
##   5 | 02
```

```
stem(all_data$C, scale = 1)
```

```
##
##   The decimal point is 1 digit(s) to the right of the |
##
##   2 | 4
##   2 | 56799
##   3 | 1334444
##   3 | 5666779999
##   4 | 000122344
##   4 | 556
##   5 | 0
```

```
stem(all_data$D, scale = 1)
```

```
##
##   The decimal point is at the |
##
##   28 | 57
##   30 | 58
##   32 | 5791556
##   34 | 157
##   36 | 69905
##   38 | 3457712
##   40 | 4662688
```

```
##   42 | 33
##   44 |
##   46 | 5
```

Stripchart

```
stripchart(all_data$A)
```

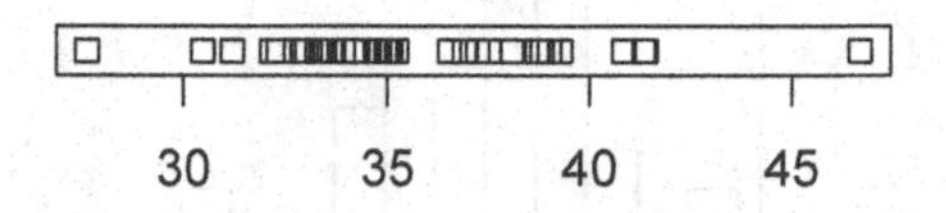


```
stripchart(all_data$B)
```

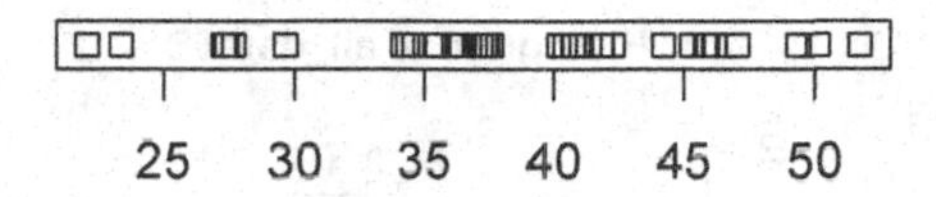


```
stripchart(all_data$C)
```

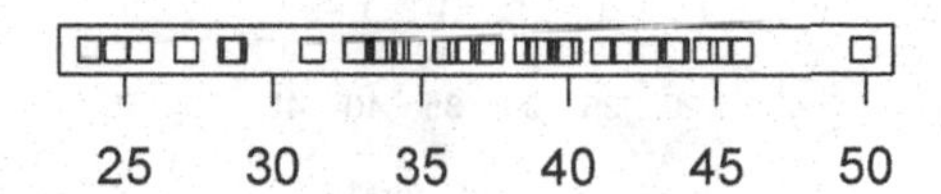


```
stripchart(all_data$D)
```

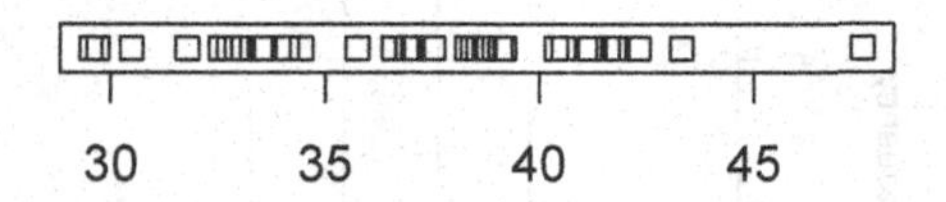

Histograms

```
hist(all_data$A)
```

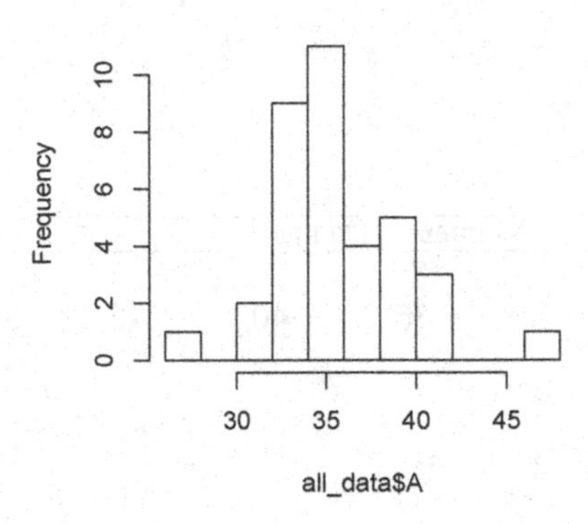


```
hist(all_data$B)
```

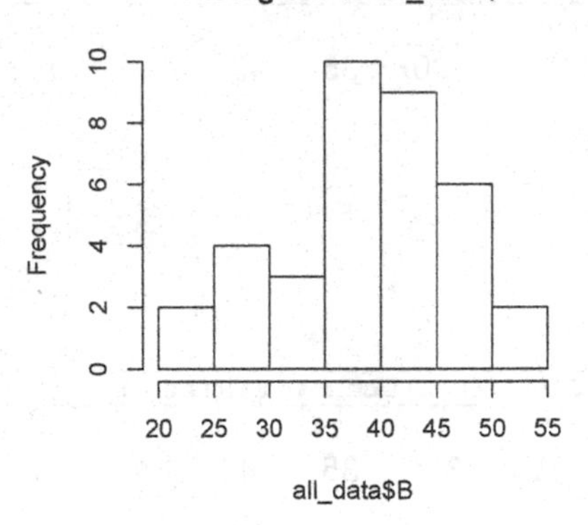


```
hist(all_data$C)
```

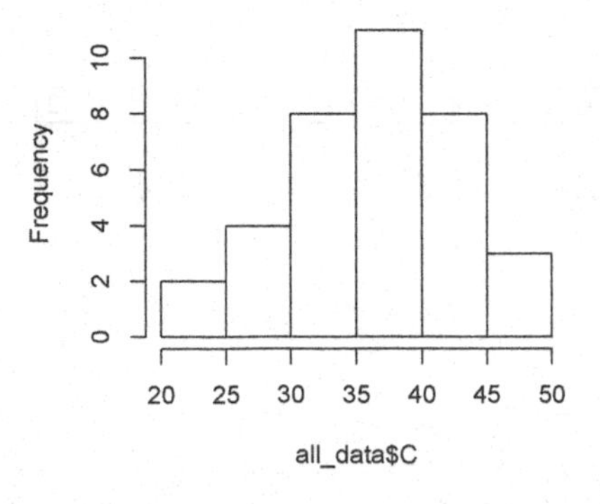

```
hist(all_data$D)
```

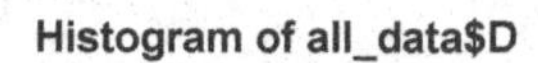

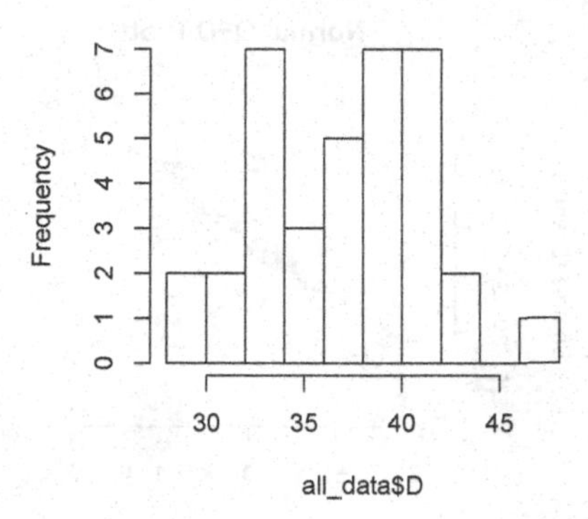

Boxplot of all data

```
boxplot(all_data)
```

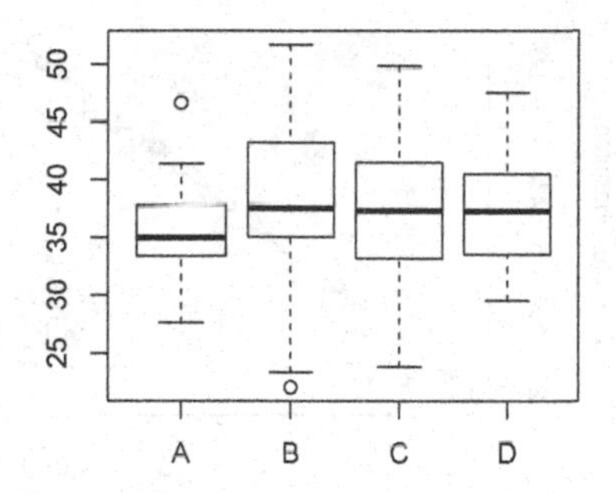

In order to prepare a qqplot

```
qqnorm(all_data$A, pch = 1, frame = FALSE)
qqline(all_data$A, col = "steelblue", lwd = 2)
```

Normal Q-Q Plot

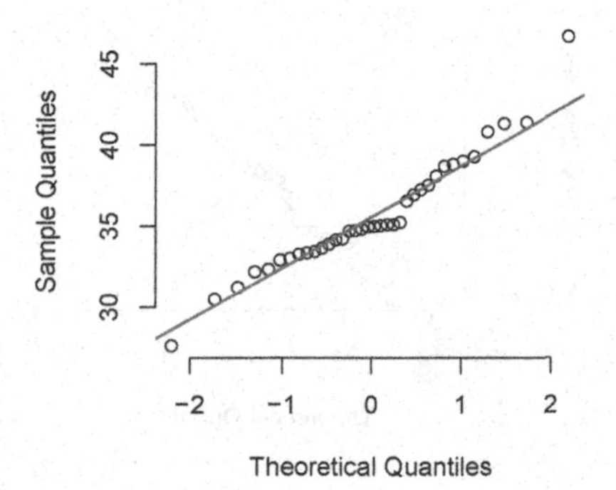

```
qqnorm(all_data$B, pch = 1, frame = FALSE)
qqline(all_data$B, col = "steelblue", lwd = 2)
```

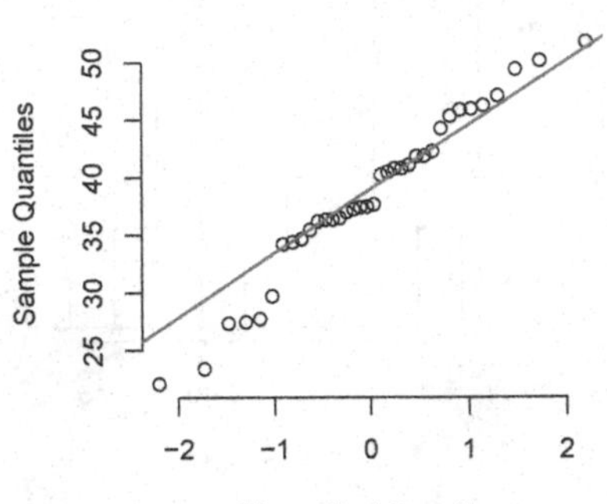


```
qqnorm(all_data$C, pch = 1, frame = FALSE)
qqline(all_data$C, col = "steelblue", lwd = 2)
```

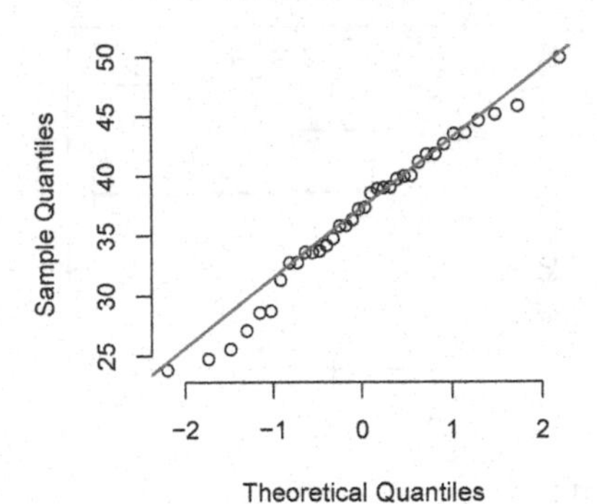


```
qqnorm(all_data$D, pch = 1, frame = FALSE)
qqline(all_data$D, col = "steelblue", lwd = 2)
```

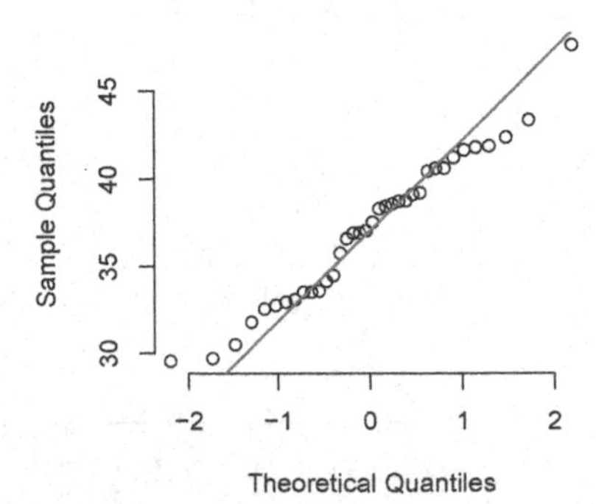

To calculate the z-scores

```
za <- (all_data$A - mean(all_data$A)) /
  sd(all_data$A)
zb <- (all_data$B - mean(all_data$B)) /
  sd(all_data$B)
zc <- (all_data$C - mean(all_data$C)) /
  sd(all_data$C)
zd <- (all_data$D - mean(all_data$D)) /
  sd(all_data$D)

all_zscores <- data.frame(za,zb,zc,zd)
```

Example of a paired t-test

```
library(broom)
library(purrr)
AB <- t.test(all_data$A, all_data$B)
```

Preparing the data-set for ANOVA

```
all_melt<- data.frame(melt(all_data))
```

```
## No id variables; using all as measure variables
```

```
lab.mod = aov(value ~ variable, data = all_melt)
summary(lab.mod)
```

```
##              Df Sum Sq Mean Sq F value Pr(>F)
## variable      3    140   46.77   1.518  0.212
## Residuals   140   4312   30.80
```

Tukey Test

```
tuk <- TukeyHSD(lab.mod)
tuk
```

```
##   Tukey multiple comparisons of means
##     95% family-wise confidence level
##
## Fit: aov(formula = value ~ variable, data = all_melt)
##
## $variable
##           diff        lwr       upr     p adj
```

```
## B-A  2.7690533 -0.6323383 6.170445 0.1528190
## C-A  1.1133456 -2.2880460 4.514737 0.8298269
## D-A  1.4339142 -1.9674774 4.835306 0.6924817
## C-B -1.6557077 -5.0570994 1.745684 0.5862584
## D-B -1.3351391 -4.7365308 2.066253 0.7376566
## D-C  0.3205686 -3.0808230 3.721960 0.9948066
```

In order to show another example of anova with different models we load the other data-set

```
load(file = "myDrive/MyPath/Hu_anova.RData")
```

Considering the independent variables

```
res.aov2 <- aov(value ~ label.a +
                  label.conc, data = hu_melt)
summary(res.aov2)
```

```
##             Df Sum Sq Mean Sq F value   Pr(>F)
## label.a      1    255     255   17.28 0.000109 ***
## label.conc   1  10265   10265  696.52  < 2e-16 ***
## Residuals   57    840      15
## ---
## Signif. codes:  0 '***' 0.001 '**' 0.01 '*' 0.05 '.' 0.1 ' ' 1
```

```
multg <- summary(glht(res.aov2,
                      linfct = mcp(label.conc = "Tukey")))
```

Two way anova with interaction

```
res.aov3 <- aov(value ~ label.a *label.conc,
                data =  hu_melt)
summary(res.aov3)
```

```
##                     Df Sum Sq Mean Sq F value   Pr(>F)
## label.a              1    255     255  18.560 6.71e-05 ***
## label.conc           1  10265   10265 747.968  < 2e-16 ***
## label.a:label.conc   1     72      72   5.211   0.0263 *
## Residuals           56    768      14
## ---
## Signif. codes:  0 '***' 0.001 '**' 0.01 '*' 0.05 '.' 0.1 ' ' 1
```

```
tuc.aov3 <- TukeyHSD(res.aov3)
tuc.aov3
```

```
##   Tukey multiple comparisons of means
##     95% family-wise confidence level
##
## Fit: aov(formula = value ~ label.a * label.conc, data = hu_melt)
##
## $label.a
##              diff       lwr      upr    p adj
## GO-FGO -4.120715 -6.036801 -2.20463 6.71e-05
##
## $label.conc
##               diff      lwr      upr p adj
## 0.5-0.25 26.15914 24.24305 28.07522     0
##
## $`label.a:label.conc`
##                          diff       lwr       upr     p adj
## GO:0.25-FGO:0.25 -1.937345 -5.519104  1.644413 0.4849724
## FGO:0.5-FGO:0.25 28.342506 24.760748 31.924265 0.0000000
## GO:0.5-FGO:0.25  22.038421 18.456662 25.620179 0.0000000
## FGO:0.5-GO:0.25  30.279852 26.698093 33.861610 0.0000000
## GO:0.5-GO:0.25   23.975766 20.394007 27.557524 0.0000000
## GO:0.5-FGO:0.5   -6.304086 -9.885844 -2.722327 0.0001150
```

```
par(mfrow=c(2,2))
plot(res.aov3)
```

```
aov_residuals <- residuals(object = res.aov3)
```

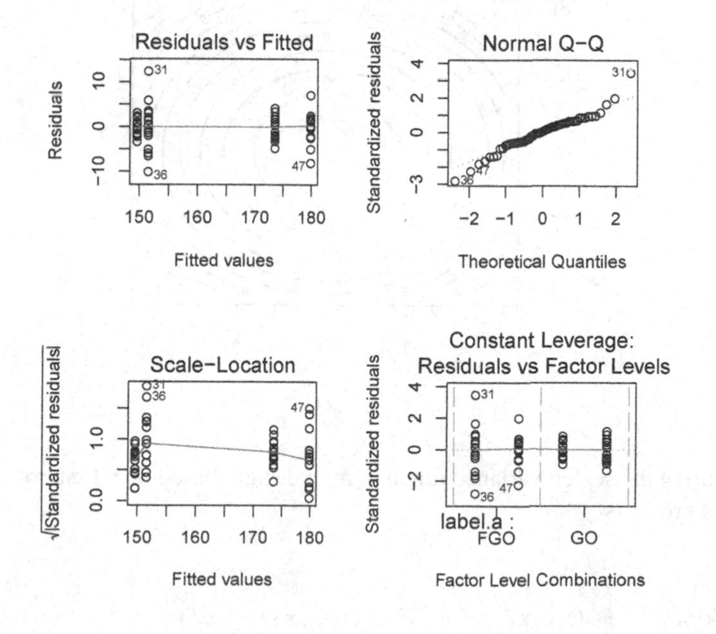

A.3 CHAPTER 2

To load the data-set

```
load(file =  "myDrive/MyPath/Emagma.RData")
```

Create the coded data

```
doe <- data.frame(EMAGMA,CE,Screw.speed,is)
doe$x1 <- (doe$EMAGMA-mean(doe$EMAGMA))/5
doe$x2 <- (doe$CE-mean(doe$CE))/0.25
doe$x3 <- (doe$Screw.speed-mean(doe$Screw.speed))/50
doe11 <- doe[doe$EMAGMA!=15,]
doe22 <- doe11[doe11$CE!=0.25,]
```

Computing the scaled variance function for a design, based on a first order model

```
par(mfrow=c(1,1))
varfcn(doe, ~ FO(x1,x2,x3), contour = TRUE)
```

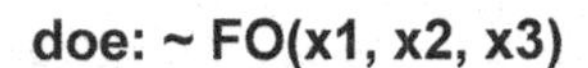

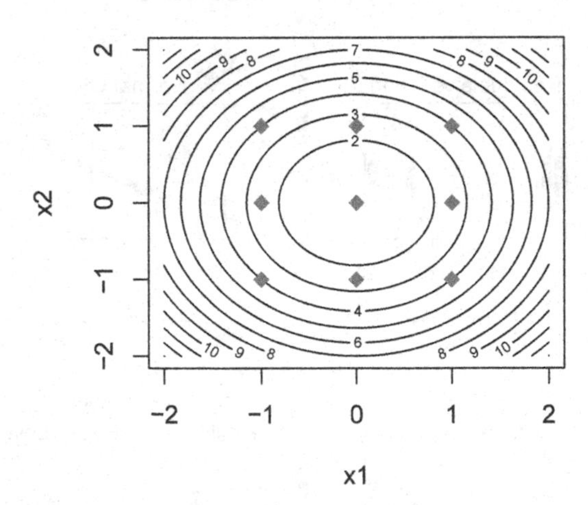

Computing the scaled variance function for a design, based on a first order model with cross products

```
par(mfrow=c(1,1))
varfcn(doe, ~ FO(x1,x2,x3)+TWI(x1,x2,x3), contour = TRUE)
```

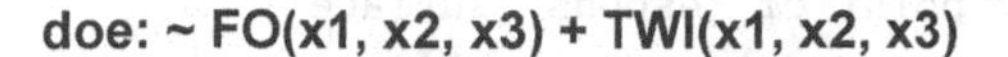

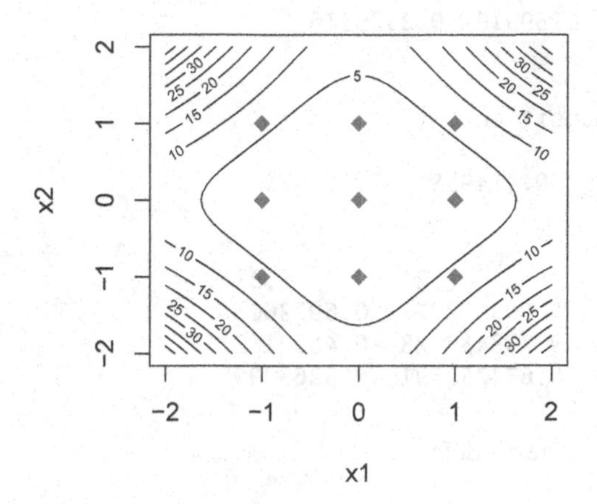

Computing the model and reporting the summary

```
mod = rsm(is ~ FO(x1,x2,x3)+TWI(x1,x2,x3), data=doe)
summary(mod)
```

```
##
## Call:
## rsm(formula = is ~ FO(x1, x2, x3) + TWI(x1, x2, x3), data = doe)
##
##             Estimate Std. Error t value  Pr(>|t|)
## (Intercept) 93.55556    3.93603 23.7690 8.325e-11 ***
## x1          43.91667    4.82064  9.1101 1.861e-06 ***
## x2          20.16667    4.82064  4.1834 0.0015278 **
## x3          18.00000    3.93603  4.5731 0.0007993 ***
## x1:x2       10.12500    5.90405  1.7149 0.1143561
## x1:x3       11.08333    4.82064  2.2991 0.0420953 *
## x2:x3       -0.83333    4.82064 -0.1729 0.8658950
## ---
## Signif. codes:  0 '***' 0.001 '**' 0.01 '*' 0.05 '.' 0.1 ' ' 1
##
## Multiple R-squared:  0.9218, Adjusted R-squared:  0.8791
## F-statistic: 21.61 on 6 and 11 DF,  p-value: 1.738e-05
##
## Analysis of Variance Table
##
## Response: is
##                 Df Sum Sq Mean Sq F value    Pr(>F)
## FO(x1, x2, x3)   3  33856 11285.5 40.4697 3.103e-06
## TWI(x1, x2, x3)  3   2303   767.5  2.7523   0.09301
## Residuals       11   3067   278.9
## Lack of fit     11   3067   278.9
## Pure error       0      0
```

```
##
## Stationary point of response surface:
##         x1         x2         x3
## -1.9709768 -4.6139918  0.2526316
##
## Eigenanalysis:
## eigen() decomposition
## $values
## [1]  7.3014119  0.4149581 -7.7163701
##
## $vectors
##          [,1]         [,2]       [,3]
## x1 -0.7167997 -0.005037277  0.6972609
## x2 -0.4674754 -0.738483628 -0.4859102
## x3 -0.5173634  0.674252591 -0.5269900
```

Diagnostic plot for the model

```
par(mfrow=c(2,2))
plot(mod)
```

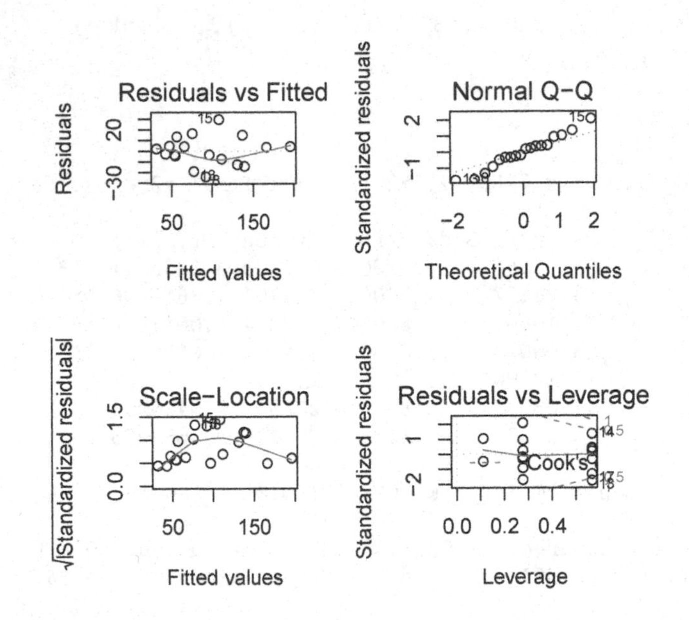

If you need a full factorial design you can rely on this syntax is for a full Factorial n=2 2 levels

```
expand.grid(A = c("+", "-"), B = c("+", "-"))
```

```
##   A B
## 1 + +
```

```
## 2 - +
## 3 + -
## 4 - -
```

of full Factorial n=3 with two levels

```
expand.grid(A = c("+", "-"),
            B = c("+", "-"), C = c("+", "-"))
```

```
##   A B C
## 1 + + +
## 2 - + +
## 3 + - +
## 4 - - +
## 5 + + -
## 6 - + -
## 7 + - -
## 8 - - -
```

For fractional factorial you can use the command

```
FrF2(nruns=8,nfactors=4)
```

```
##    A  B  C  D
## 1  1 -1 -1  1
## 2  1  1  1  1
## 3 -1 -1 -1 -1
## 4 -1  1 -1  1
## 5 -1 -1  1  1
## 6  1 -1  1 -1
## 7 -1  1  1 -1
## 8  1  1 -1 -1
## class=design, type= FrF2
```

And forPlackett-Burman

```
pb(8)
```

For the Data-set OPT CABLES

```
#setup lower (min = minus, "-") and higher(plus, "+")
#and zero, for the factor
Amin = 0.025
Aplus = 0.050
```

```
Bmin = 170
Bplus  = 190
Azero  = 0.0375
Bzero = 180
stepB=10
stepA=0.0125

#response
p1 <- 1130
a <- 1078
b <- 1398
ab <- 1320
#
A0<- c(Amin,Aplus,Amin,Aplus)
B0 <- c(Bmin,Bmin,Bplus,Bplus)
y0 <- c(p1, a,b,ab)
block0 <- rep(1,length(A0))
doe0 <- data.frame(A0,B0,y0,block0)
colnames(doe0) <- c("A","B","y","block")

sumcoeff=4 #sum of the coefficients of the model
rep=1 #number of replicate
Ahat <-  (a+ab)/2 - (p1+b)/2
Bhat <-  (b+ab)/2 - (p1+a)/2
ABhat <- (ab+p1)/2 -(a+b)/2
b1=Ahat/2
b2=Bhat/2
b12=ABhat/2
AC <- a+ab-p1-b
BC <- b+ab-p1-a
ABC <- ab+p1-a-b
#calculation of the sum of squares
SSA=(Ahat^2)/(sumcoeff*rep)
SSB=(Bhat^2)/(sumcoeff*rep)
SSAB=(ABhat)^2/(sumcoeff*rep)
```

Summary model 4 points data-set CABLE OPT

```
coded.doe0 <- coded.data(doe0,
                         x1~(B-Bzero)/stepB,
                         x2~(A-Azero)/stepA)
mod0 = rsm(y ~(FO(x1,x2)+TWI(x1,x2)), data=coded.doe0)
summary(mod0)
```

```
## Warning in anova.lm(object): ANOVA F-tests on an essentially
   perfect fit are
## an essentially perfect fit are unreliable

##
## Call:
## rsm(formula = y ~ (FO(x1, x2) + TWI(x1, x2)), data = coded.doe0)
##
##             Estimate Std. Error t value Pr(>|t|)
## (Intercept)   1231.5         NA      NA       NA
## x1             127.5         NA      NA       NA
## x2             -32.5         NA      NA       NA
## x1:x2           -6.5         NA      NA       NA
##
## Multiple R-squared:       1,  Adjusted R-squared:    NaN
## F-statistic:   NaN on 3 and 0 DF,  p-value: NA
##
## Analysis of Variance Table
##
## Response: y
##             Df Sum Sq Mean Sq F value Pr(>F)
## FO(x1, x2)   2  69250   34625
## TWI(x1, x2)  1    169     169
## Residuals    0      0
## Lack of fit  0      0
## Pure error   0      0
##
## Stationary point of response surface:
##       x1       x2
## -5.00000 19.61538
##
## Stationary point in original units:
##           B           A
## 130.0000000   0.2826923
##
## Eigenanalysis:
## eigen() decomposition
## $values
## [1]  3.25 -3.25
##
## $vectors
##          [,1]       [,2]
## x1 -0.7071068 -0.7071068
## x2  0.7071068 -0.7071068
```

Adding one central point

```
p0=c(1205,1200,1195)
A1<- c(Amin,Aplus,Azero,Azero,Azero,Amin,Aplus)
B1 <- c(Bmin,Bmin,Bzero,Bzero,Bzero,Bplus,Bplus)
y1 <- c(p1, a, p0,b,ab)
block1 <- rep(1,length(A1))
doe1 <- data.frame(A1,B1,y1,block1)
colnames(doe1) <- c("A","B","y","block")
coded.doe1 <- coded.data(doe1, x1~(B-Bzero)/stepB,
                         x2~(A-Azero)/stepA)
mod1 = rsm(y ~(FO(x1,x2)+TWI(x1,x2)), data=coded.doe1)
summary(mod1)
```

```
##
## Call:
## rsm(formula = y ~ (FO(x1, x2) + TWI(x1, x2)), data = coded.doe1)
##
##              Estimate Std. Error  t value  Pr(>|t|)
## (Intercept) 1218.0000     9.1313 133.3871 9.291e-07 ***
## x1           127.5000    12.0796  10.5550  0.001817 **
## x2           -32.5000    12.0796  -2.6905  0.074383 .
## x1:x2         -6.5000    12.0796  -0.5381  0.627869
## ---
## Signif. codes:  0 '***' 0.001 '**' 0.01 '*' 0.05 '.' 0.1 ' ' 1
##
## Multiple R-squared:  0.9754, Adjusted R-squared:  0.9508
## F-statistic: 39.65 on 3 and 3 DF,  p-value: 0.006503
##
## Analysis of Variance Table
##
## Response: y
##              Df Sum Sq Mean Sq F value   Pr(>F)
## FO(x1, x2)   2  69250   34625 59.3232 0.003873
## TWI(x1, x2)  1    169     169  0.2895 0.627869
## Residuals    3   1751     584
## Lack of fit  1   1701    1701 68.0400 0.014381
## Pure error   2     50      25
##
## Stationary point of response surface:
##        x1       x2
## -5.00000 19.61538
##
## Stationary point in original units:
##           B           A
## 130.0000000   0.2826923
##
## Eigenanalysis:
```

```
## eigen() decomposition
## $values
## [1]  3.25 -3.25
##
## $vectors
##            [,1]       [,2]
## x1 -0.7071068 -0.7071068
## x2  0.7071068 -0.7071068
```

Diagnostic plots for model with central points for the data-set

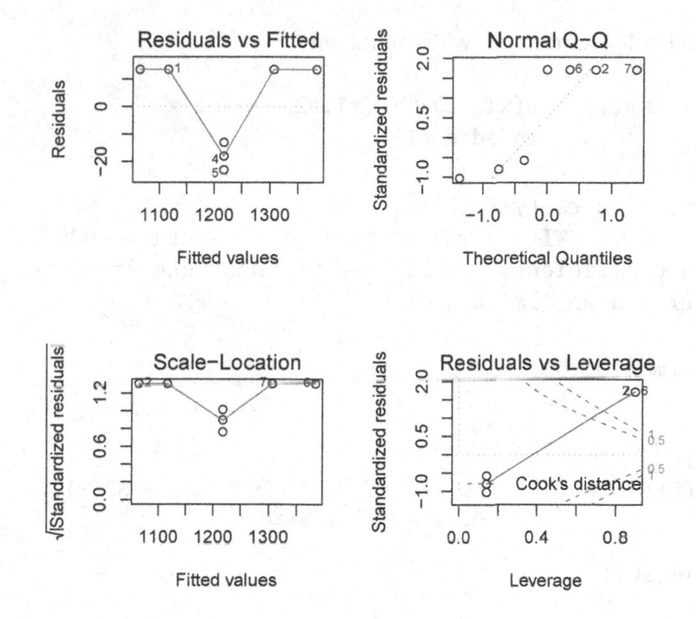

Plot of the interactions for model with central points

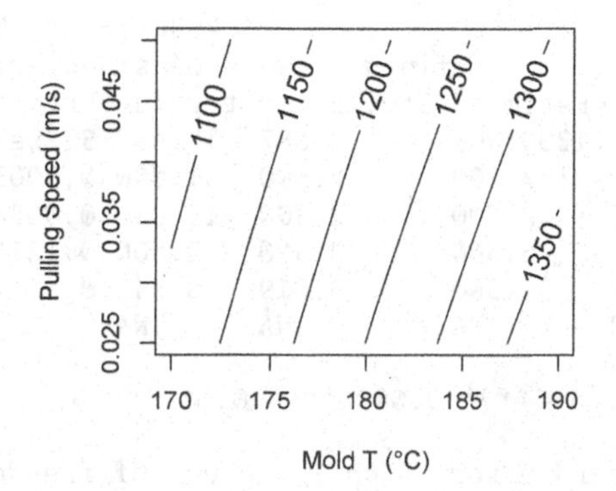

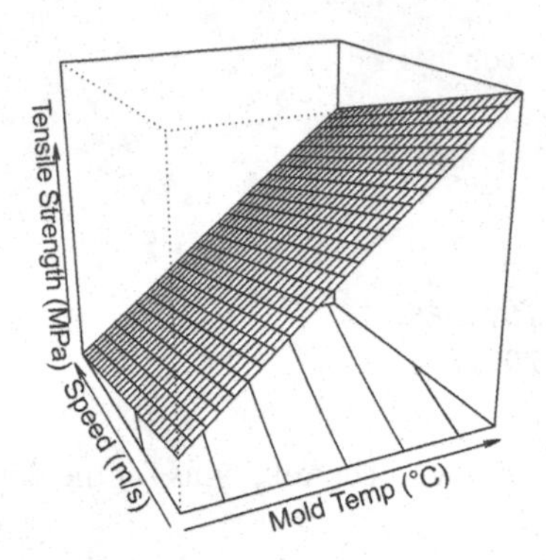

Second order for data-set with central point

```
mod1q = rsm(y ~(FO(x1,x2)+TWI(x1,x2)+SO(x1,x2)),
            data=coded.doe1)
```

```
## Warning in rsm(y ~ (FO(x1, x2) +
##             TWI(x1, x2) + SO(x1, x2)), data = coded.doe1):
## Some coefficients are aliased - cannot use 'rsm' methods.
## Returning an 'lm' object.
```

```
summary(mod1q)
```

```
##
## Call:
## rsm(formula = y ~ (FO(x1, x2) + TWI(x1, x2) + SO(x1, x2)),
##                     data = coded.doe1)
##
## Residuals:
##           1           2           3           4           5
##  8.568e-16  4.678e-16  5.000e+00  5.763e-15 -5.000e+00
##
##                                               6           7
##                                       5.703e-16   5.703e-16
## Coefficients: (1 not defined because of singularities)
##                  Estimate Std. Error t value Pr(>|t|)
## (Intercept)      1200.000      2.887 415.692 5.79e-06 ***
## FO(x1, x2)x1      127.500      2.500  51.000 0.000384 ***
## FO(x1, x2)x2      -32.500      2.500 -13.000 0.005865 **
## TWI(x1, x2)        -6.500      2.500  -2.600 0.121541
## PQ(x1, x2)x1^2     31.500      3.819   8.249 0.014381 *
## PQ(x1, x2)x2^2         NA         NA      NA       NA
## ---
## Signif. codes:  0 '***' 0.001 '**' 0.01 '*' 0.05 '.' 0.1 ' ' 1
##
## Residual standard error: 5 on 2 degrees of freedom
## Multiple R-squared:  0.9993, Adjusted R-squared:  0.9979
## F-statistic: 711.2 on 4 and 2 DF,  p-value: 0.001405
```

Data-set Faria with replicates

```
load(file = "myDrive/MyPath/Faria.RData")
mod3 = rsm(y ~(FO(x1,x2)+TWI(x1,x2)+SO(x1,x2)),
        data=coded.doe3)
```

Plot of the interactions Faria

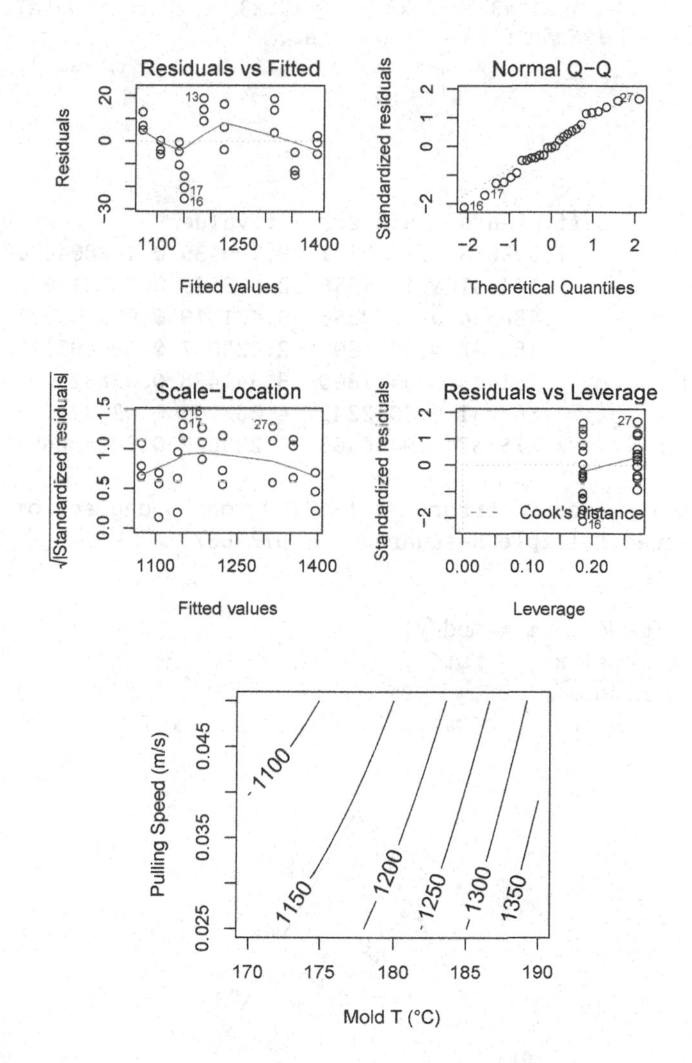

Mixture models
#Mixture Data

```
x1 <- c(0,1,0,0.167,0.5,0.0,0.333,0.167,0.5,0.667)
x2 <- c(1,0,0,0.667,0.000,0.500,0.333,0.167,0.500,0.167)
```

```
x3 <- c(0.00,0.00,1.00,0.167,0.500,0.500,0.333,0.667,0.000,0.167)
y1 <- c(2.12,1.71,3.46,3.05,3.41,3.72,3.21,3.74,2.48,2.35)
y2 <- c(3.14,1.02,3.64,3.96,3.23,4.61,4.23,4.48,3.51,2.62)

my.data <- data.frame(x1,x2,x3,y1,y2)
my.data2 <- data.frame(x1,x2,x3,y1,y2)

lm.mod.y1 <- lm(y1 ~ -1 + x1 + x2 + x3 + x1:x2
                + x1:x3 + x2:x3 + x1:x2:x3, data = my.data)
mix.mod.y1<- MixModel(frame = my.data2,
                "y1", mixcomps = c("x1", "x2", "x3"),
                model = 4)
```

```
##
##            coefficients   Std.err   t.value         Prob
## x1            1.629618 0.1790881  9.099535 0.0028044408
## x2            2.164831 0.1790656 12.089599 0.0012179800
## x3            3.486624 0.1790656 19.471210 0.0002959262
## x2:x1         2.188107 0.9011800  2.428047 0.0934931752
## x3:x1         3.191473 0.9011800  3.541438 0.0383242918
## x2:x3         3.863731 0.9012242  4.287203 0.0233245326
## x2:x3:x1     -7.275187 5.9456403 -1.223617 0.3084366979
##
## Residual standard error:  0.1852119  on  3 degrees of freedom
## Corrected Multiple R-squared:  0.9774687
```

```
ModelPlot(model = mix.mod.y1,
        dimensions = list(x1 = "x1", x2 = "x2", x3 = "x3"),
        contour = TRUE,  axislabs =  c("x1", "x2", "x3"),
        cornerlabs = c("A", "B", "C") ,pseudo=F)
```

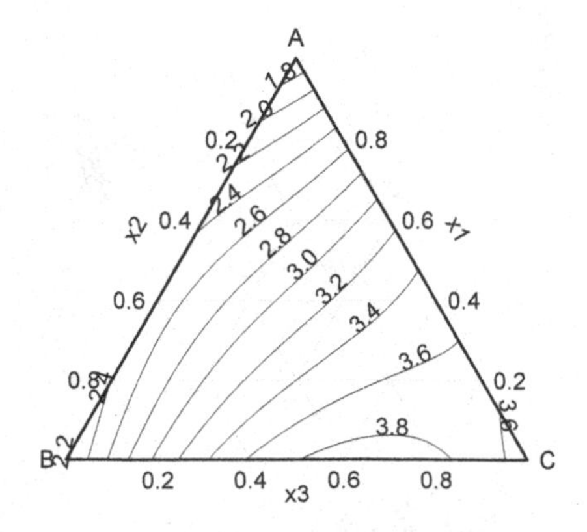

```
lm.mod.y2 <- lm(y2 ~ -1 + x1 + x2 + x3 + x1:x2 +
                 x1:x3 + x2:x3 +
                 x1:x2:x3, data = my.data)
mix.mod.y2<- MixModel(frame = my.data2, "y2",
                      mixcomps = c("x1", "x2", "x3"),
                      model = 4)
```

```
##
##           coefficients   Std.err     t.value         Prob
## x1           0.9369177 0.2460864  3.80727130 0.0318469238
## x2           3.1037951 0.2460554 12.61420989 0.0010743631
## x3           3.7255655 0.2460554 15.14116240 0.0006254824
## x2:x1        5.4786127 1.2383186  4.42423510 0.0214455536
## x3:x1        3.6036163 1.2383186  2.91008810 0.0619921997
## x2:x3        4.9791754 1.2383794  4.02071895 0.0276320351
## x2:x3:x1    -0.5417128 8.1699516 -0.06630551 0.9513060298
##
## Residual standard error:  0.2545012  on  3 degrees of freedom
## Corrected Multiple R-squared:  0.9806705
```

```
ModelPlot(model = mix.mod.y2,
        dimensions = list(x1 = "x1", x2 = "x2", x3 = "x3"),
        contour = TRUE,  axislabs =  c("x1", "x2", "x3"),
        cornerlabs = c("A", "B", "C") ,pseudo=F)
```

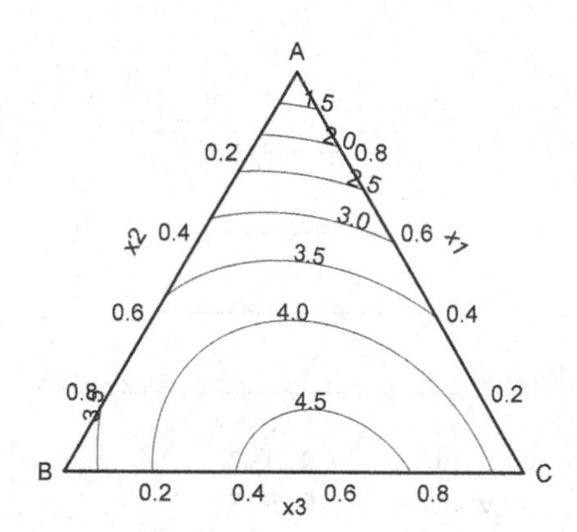


```
p1 <- c(0.333,0.25,0.143,0.2915,0.2385,0.1965,0.242,0.219)
p2 <- c(0.333,0.75,0,0.5415,0.1665,0.375,0.361,0.329)
p3 <- c(0.333,0,0.857,0.167,0.595,0.4285,0.397,0.452)
y1 <- c(9.810,22.070,7.579,15.743,10.770,11.593,15.509,12.095)
orig <-data.frame(p1,p2,p3,y1)
```

```
title <- c("Actual Component Space", "Pseudo Component Space")
option <- c(FALSE, TRUE)

quadm<- lm(y1 ~ -1 + p1 + p2 + p3
          + p1:p2 + p1:p3 + p2:p3 + p1:p2:p3, data = orig)
title<-c("Actual Component Space", "Pseudo Component Space")

 ModelPlot(model = quadm,
       dimensions = list(p1 = "p1", p2 = "p2", p3 = "p3"),
 constraints = F,  contour = TRUE, cuts= 10, fill = TRUE,
 axislabs = c("x1", "x2", "x3"),
 cornerlabs = c("x1", "x2", "x3"), pseudo = F)
```

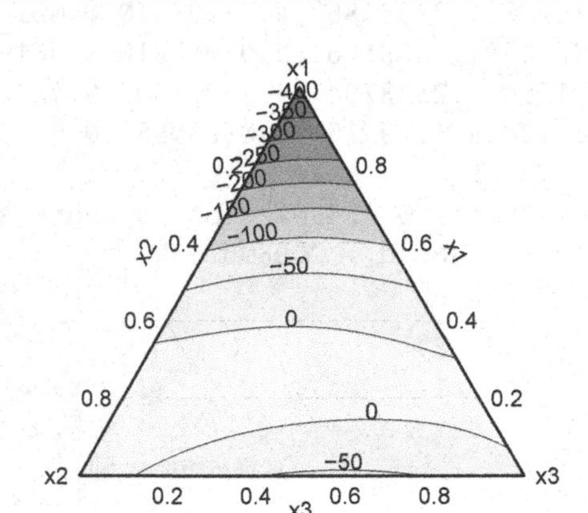

A.4 CHAPTER 3

```
library(tidyverse)
```

```
## -- Attaching packages ------------- tidyverse 1.3.0 --

## v ggplot2 3.3.0     v purrr   0.3.3
## v tibble  2.1.3     v dplyr   0.8.5
## v tidyr   1.0.2     v stringr 1.4.0
## v readr   1.3.1     v forcats 0.5.0

## -- Conflicts ----------------- tidyverse_conflicts() --
## x dplyr::filter() masks stats::filter()
## x dplyr::lag()    masks stats::lag()
```

```
library(knitr)
library(qqplotr)
```

```
##
## Attaching package: 'qqplotr'

## The following objects are masked from 'package:ggplot2':
##
##     stat_qq_line, StatQqLine
```

```
library(reshape2)
```

```
##
## Attaching package: 'reshape2'
## The following object is masked from 'package:tidyr':
##
##     smiths
```

```
library(factoextra)
```

```
## Welcome! Want to learn more?
## See two factoextra-related books at https://goo.gl/ve3WBa
```

```
library(FactoMineR)
library(caret)
```

```
## Loading required package: lattice

##
## Attaching package: 'caret'
## The following object is masked from 'package:purrr':
##
##     lift
```

```
library(multcomp)
```

```
## Loading required package: mvtnorm

## Loading required package: survival
## Attaching package: 'survival'

## The following object is masked from 'package:caret':
##
##     cluster
```

```
## Loading required package: TH.data

## Loading required package: MASS

##
## Attaching package: 'MASS'
## The following object is masked from 'package:dplyr':
##
##     select

##
## Attaching package: 'TH.data'
## The following object is masked from 'package:MASS':
##
##     geyser
```

Create a simulated data-set for showing correlation of variables

```
set.seed(123)
spurious_data <- data.frame(x = rnorm(500, 10, 1),
                            y = rnorm(500, 9, 1),
                            z = rnorm(500, 30, 6))
cor(spurious_data$x, spurious_data$y)
```

```
## [1] -0.05193691
```

```
plot(spurious_data$x, spurious_data$y)
```

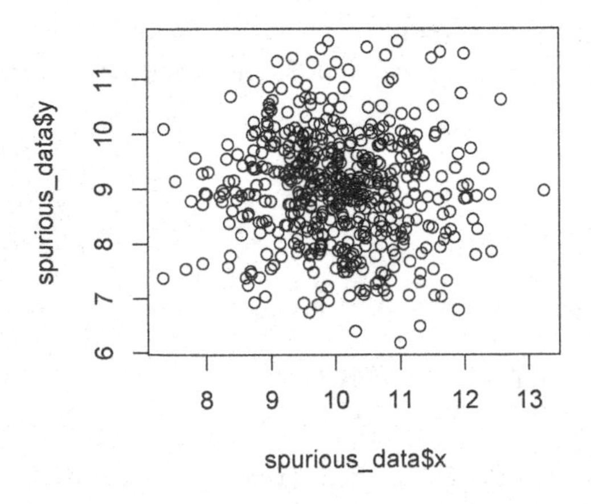

```
#same data divided for variable Z

ratio_xz <- spurious_data$x / spurious_data$z
ratio_yz <- spurious_data$y / spurious_data$z
cor(ratio_xz,ratio_yz)
```

```
## [1] 0.7986965
```

```
plot(ratio_xz,ratio_yz)
```

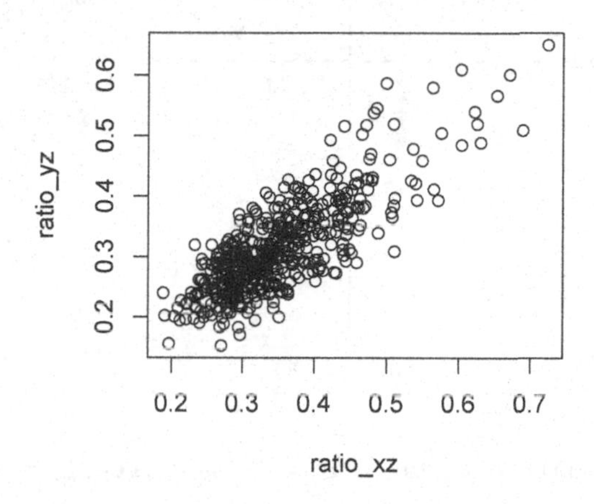


```
load(file = "myDrive/MyPath/El_extend.RData")
asphalt.pca <- PCA((Ag1[3:length(Ag1)]),
                   graph = FALSE,
                   scale.unit = FALSE)
#if you t want to scale just change scale.unit = TRUE
var2 <- get_pca_var(asphalt.pca)
print(var2$contrib,digits=3)
```

```
##       Dim.1    Dim.2    Dim.3   Dim.4    Dim.5
## Ag 6.03e+01 3.13e+01  8.10221  0.0669  0.10350
## Fe 4.85e+00 3.34e+00 86.56838  4.4022  0.05185
## Cr 5.69e-02 5.37e-02  1.94163 27.0398 28.89822
## Mn 3.55e-02 4.98e-04  2.18278 23.4792 16.22596
## Cd 2.82e-01 8.29e-02  0.00724  5.9261 20.21854
## Co 3.49e-02 2.69e-02  0.22302  3.3633  0.67831
## Pb 3.44e+01 6.50e+01  0.01749  0.1692  0.00877
## Ni 2.26e-08 1.05e-02  0.15130  5.1953 19.98146
## Zn 6.32e-03 7.78e-02  0.34014  1.3959 12.59566
## Cu 4.50e-02 2.99e-02  0.46582 28.9621  1.23773
```

```
fviz_pca_ind(asphalt.pca,
             col.ind=as.factor(marble.origin),
             repel=TRUE) + xlim(c=-5,5)+ylim(-5,5)
```

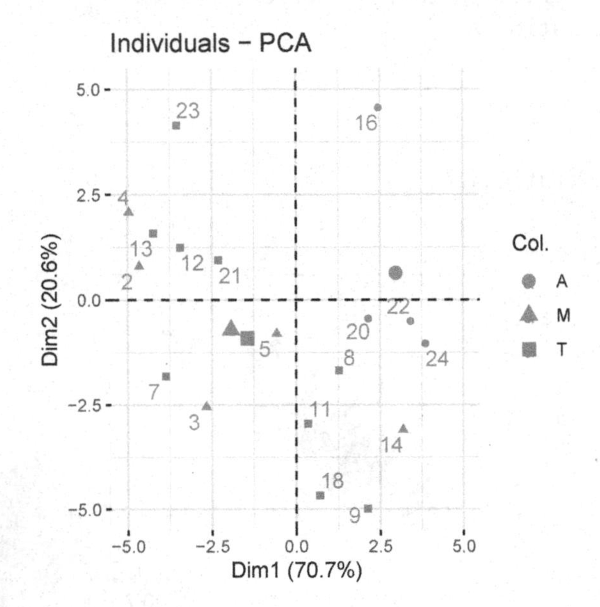


```
fviz_pca_var(asphalt.pca, col.var = "black",repel = TRUE)
```

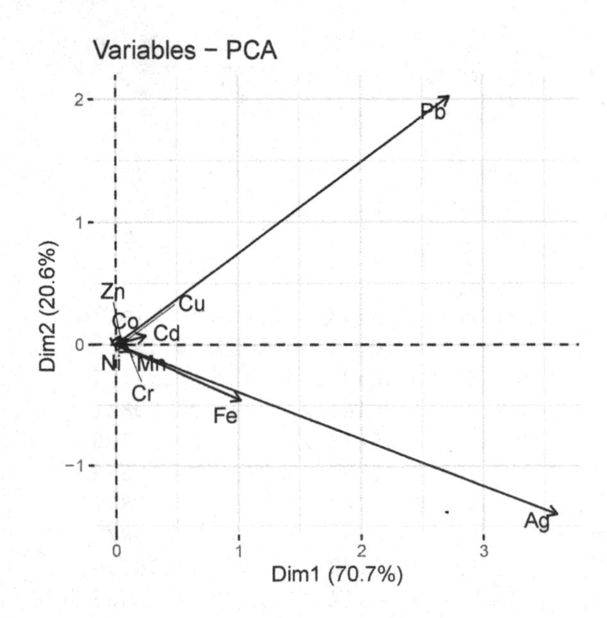

```
fviz_eig(asphalt.pca, addlabels = TRUE ,ylim = c(0, 100),
        repel = TRUE) +theme_bw()
```

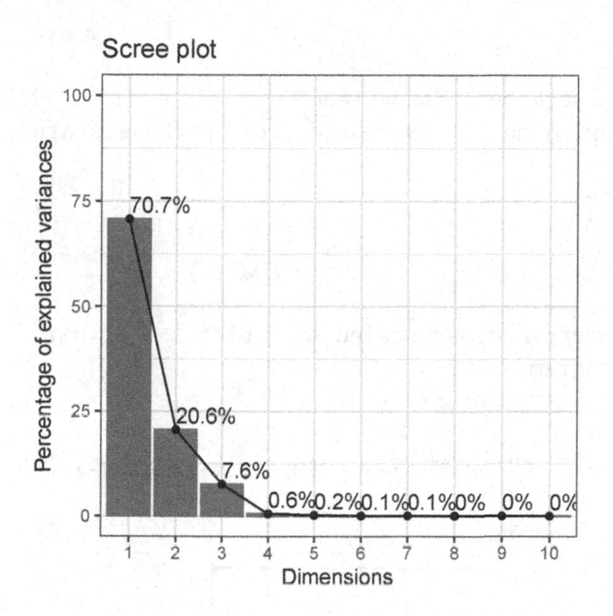


```
fviz_pca_biplot(asphalt.pca,
                col.ind=as.factor(marble.origin),
                repel = TRUE)+ xlim(c=-5,5)+ylim(-5,5)
```

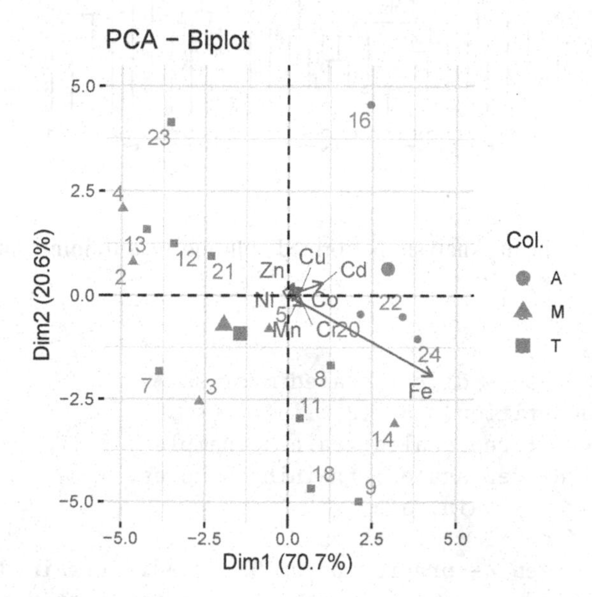

Create dendogram

```
library(dendextend)
```

```
##
## Attaching package: 'dendextend'
## The following object is masked from 'package:stats':
##
##     cutree
```

```
load(file = "myDrive/MyPath/Marble.RData")
#load data
dend <- df.norep.noclass.scaled %>%  dist %>% hclust
%>% as.dendrogram
#number of cluster choosen k=3
dend %>% plot
dend %>% rect.dendrogram(k=3, border = 8, lty = 5, lwd = 2)
```

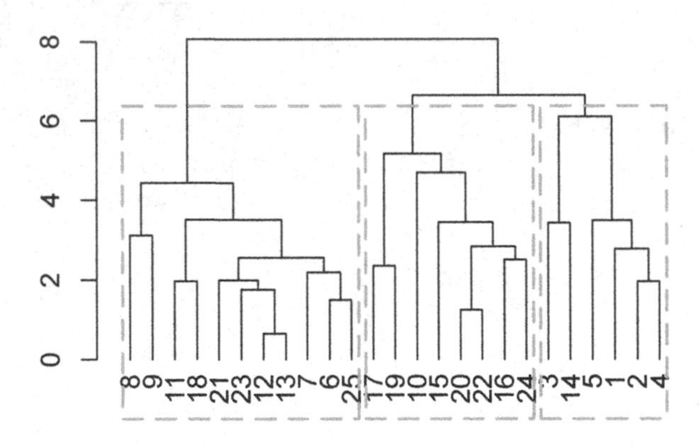

Linear Discriminant Analysis performed splitting in a training set and test set 80/20

```
set.seed(123)
training.samples <- df.rep.scaled$myclass %>%
  createDataPartition(p = 0.8, list = FALSE)
train.data <- df.rep.scaled[training.samples, ]
test.data <- df.rep.scaled[-training.samples, ]
preproc.param <- train.data %>%
  preProcess(method = c("center", "scale"))
train.transformed <- preproc.param %>% predict(train.data)
test.transformed <- preproc.param %>% predict(test.data)
```

```
model <- lda(myclass~., data = train.transformed)
# Make predictions
predictions <- model %>% predict(test.transformed)
# Model accuracy
mean(predictions$class==test.transformed$myclasses)
```

```
lda.data <- cbind(train.transformed, predict(model)$x)
model
```

```
## Call:
## lda(myclass ~ ., data = train.transformed)
##
## Prior probabilities of groups:
##         A         M         T
## 0.2407407 0.3209877 0.4382716
##
## Group means:
##            Ag         Fe          Cr         Mn         Cd         Co
## A  0.68779514  0.1015038 -0.05801378 -0.2980025  0.8990958  0.4130383
## M  0.02821522  0.7968469  0.86697858  1.1760375  0.1469056  0.7875360
## T -0.39846763 -0.6393618 -0.60310350 -0.6976317 -0.6014623 -0.8036671
##           Pb         Ni         Zn           Cu
## A  0.9505316 -0.4140419  0.1483330  0.009897638
## M  0.1063003  0.9327280  0.9544703  0.192893551
## T -0.5999767 -0.4556933 -0.7805273 -0.146710881
##
## Coefficients of linear discriminants:
##            LD1         LD2
## Ag  0.46682548  0.30409087
## Fe -0.55702818  0.03864314
## Cr  0.16342388 -0.23460888
## Mn -1.18436890 -0.56625634
## Cd -0.74974078  0.58505259
## Co -0.38953852  0.27387343
## Pb  0.47953719  0.40672299
## Ni -0.58699204 -0.15983264
## Zn -1.07245604  0.24281827
## Cu -0.02362862 -0.01825156
##
## Proportion of trace:
##    LD1    LD2
## 0.9037 0.0963
```

MDA modellig and plot

```
## Call:
## mda(formula = myclass ~ ., data = train.transformed)
##
```

```
## Dimension: 8
##
## Percent Between-Group Variance Explained:
##     v1     v2     v3     v4     v5     v6     v7     v8
##  36.49  64.69  85.05  92.73  95.94  98.59  99.69 100.00
##
## Degrees of Freedom (per dimension): 11
##
## Training Misclassification Error: 0 ( N = 162 )
##
## Deviance: 1.911
```

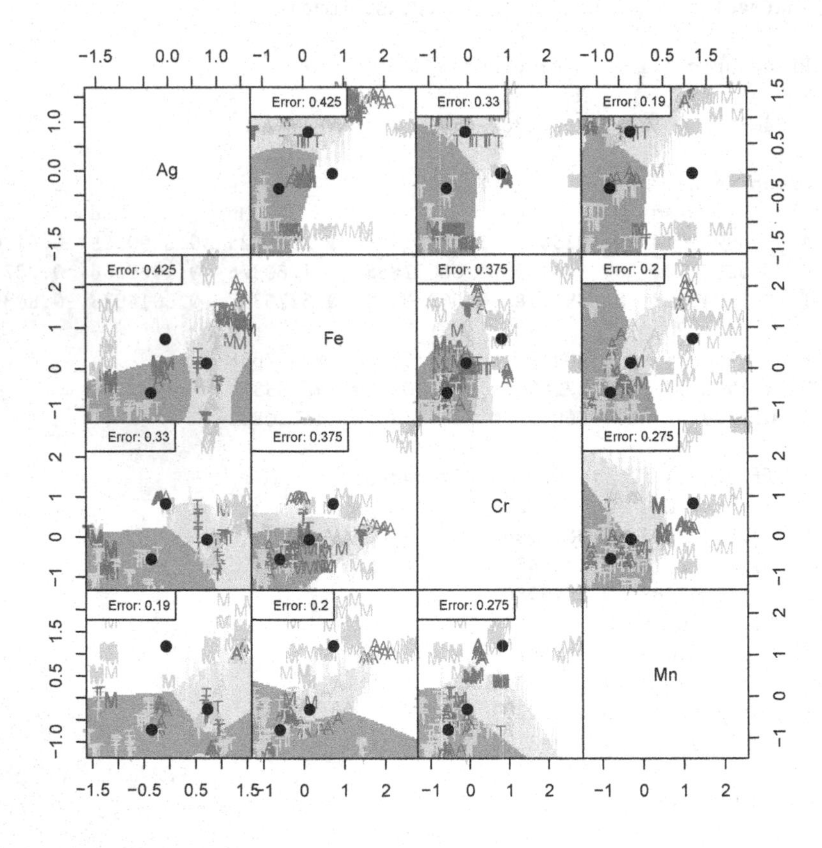

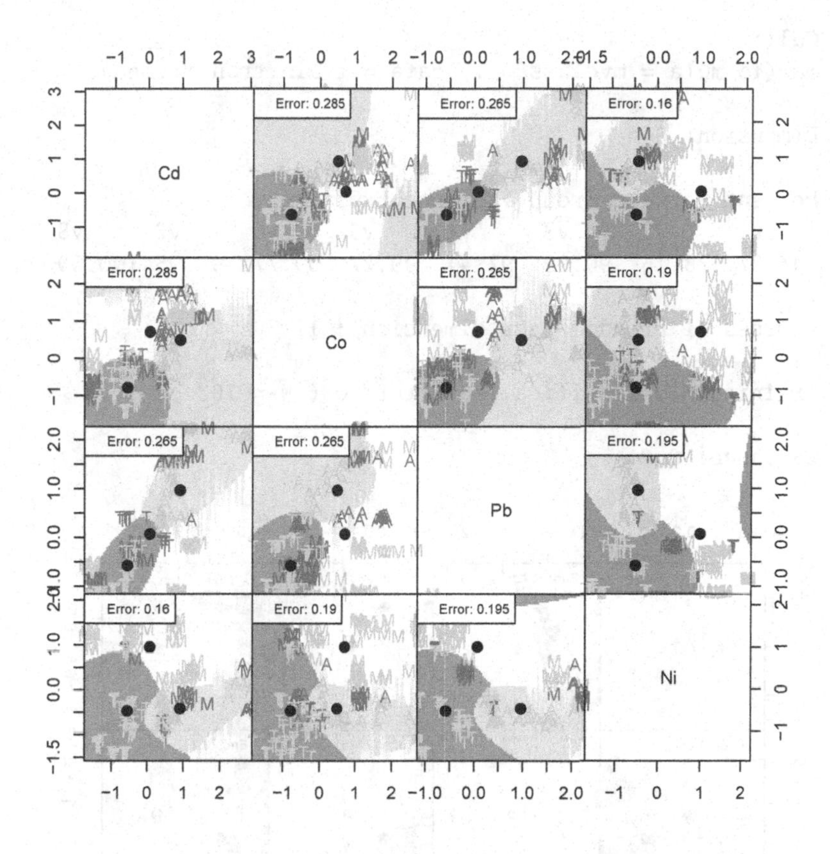
Cd
Co
Pb
Ni
Error: 0.285
Error: 0.265
Error: 0.16
Error: 0.19
Error: 0.195

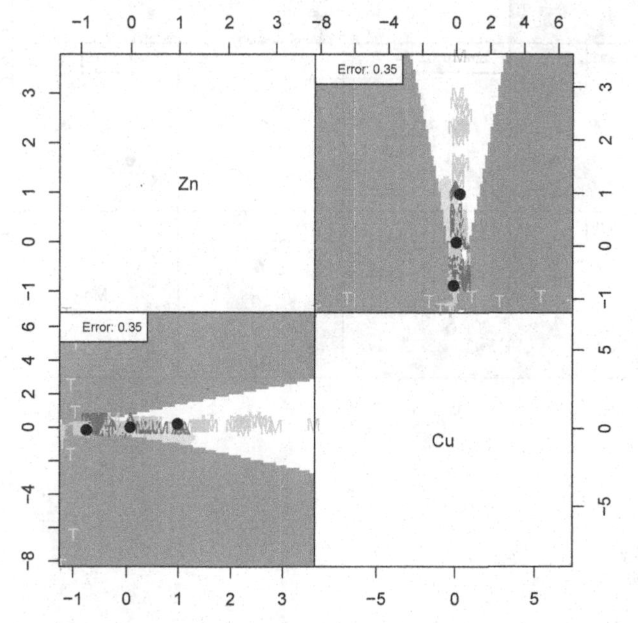
Zn
Cu
Error: 0.35

```
## Call:
## mda(formula = myclass ~ ., data = train.transformed)
##
## Dimension: 8
##
## Percent Between-Group Variance Explained:
##     v1     v2     v3     v4     v5     v6     v7     v8
##  44.97  73.16  90.99  97.41  99.27  99.77  99.95 100.00
##
## Degrees of Freedom (per dimension): 11
##
## Training Misclassification Error: 0 ( N = 162 )
##
## Deviance: 0.01
```

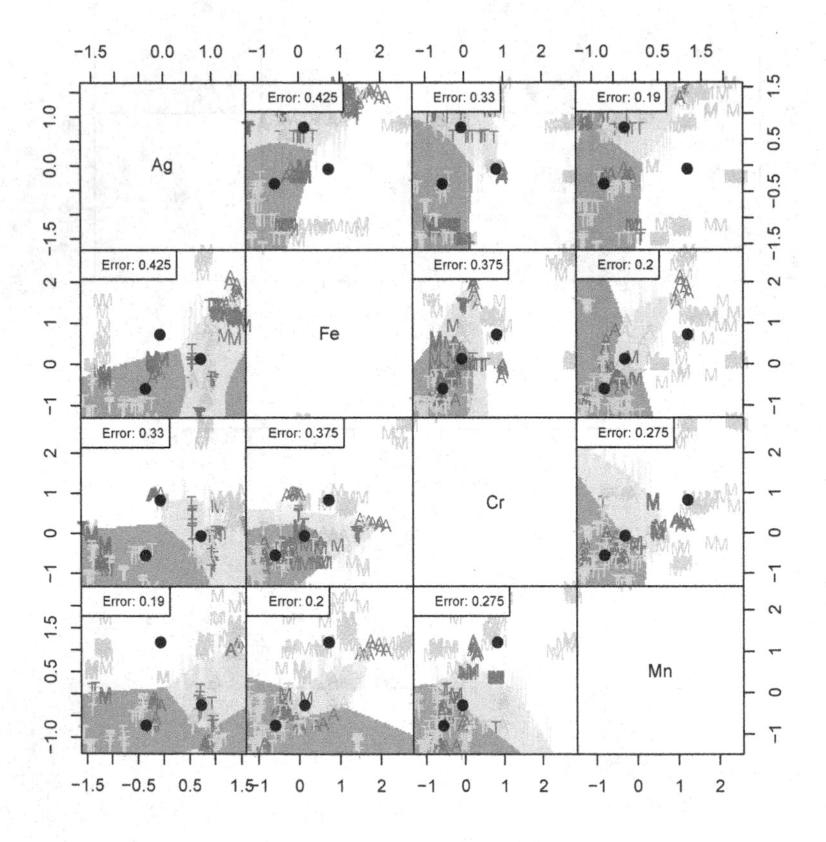

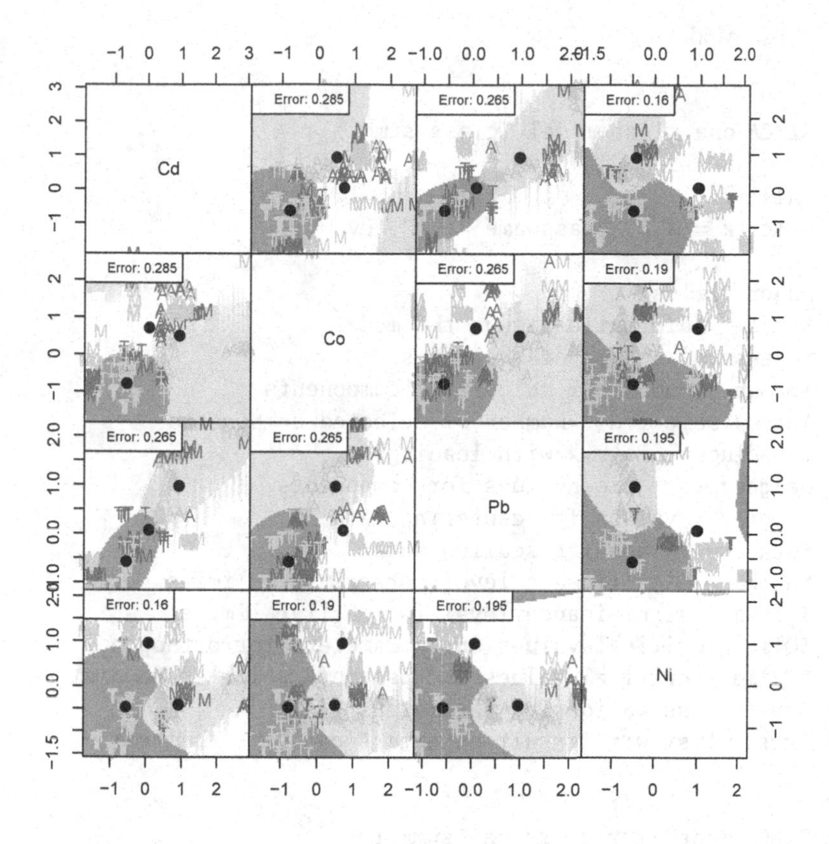
Cd
Co
Pb
Ni
Error: 0.285
Error: 0.265
Error: 0.16
Error: 0.19
Error: 0.195

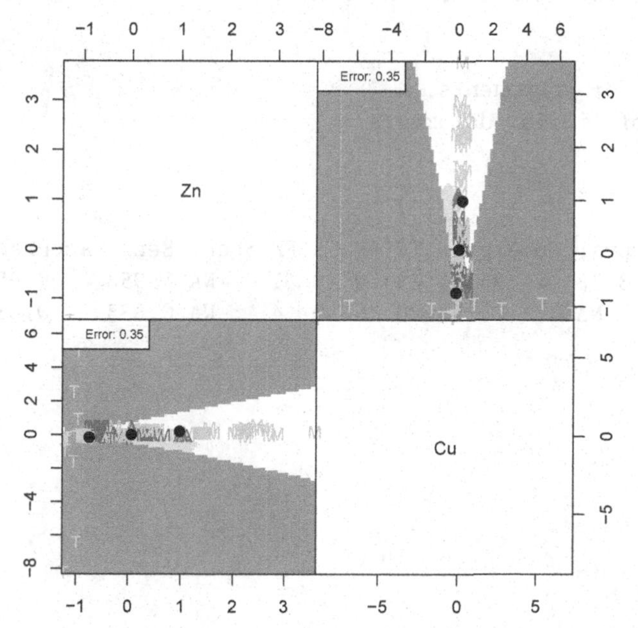
Zn
Cu
Error: 0.35

SIMCA Modelling

```
##
## SIMCA one class model (class simca)
##
## Call:
## simca(x = X.A, classname = "A", cv = 1)
##
## Major fields:
## $info - information about the model
## $classname - name of the class
## $ncomp - number of calculated components
## $ncomp.selected - number of selected components
## $loadings - matrix with loadings
## $eigenvals - eigenvalues for components
## $center - values for centering data
## $scale - values for scaling data
## $alpha - significance level for critical limits
## $gamma - significance level for outlier limits
## $Qlim - critical values and parameters for orthogonal distances
## $T2lim - critical values and parameters for score distances
## $cv - cross-validation parameters
## $res - list with result objects ('cal', 'cv', 'test'

##
## SIMCA model for class 'A' summary
##
##
## Number of components: 4
## Type of limits: ddmoments
## Alpha: 0.05
## Gamma: 0.01
##
##     Expvar Cumexpvar TP FP TN FN Spec. Sens. Accuracy
## Cal  13.38     96.25 23  0  0  1    NA 0.958    0.958
## Cv      NA        NA 20  0  0  4    NA 0.833    0.833
```

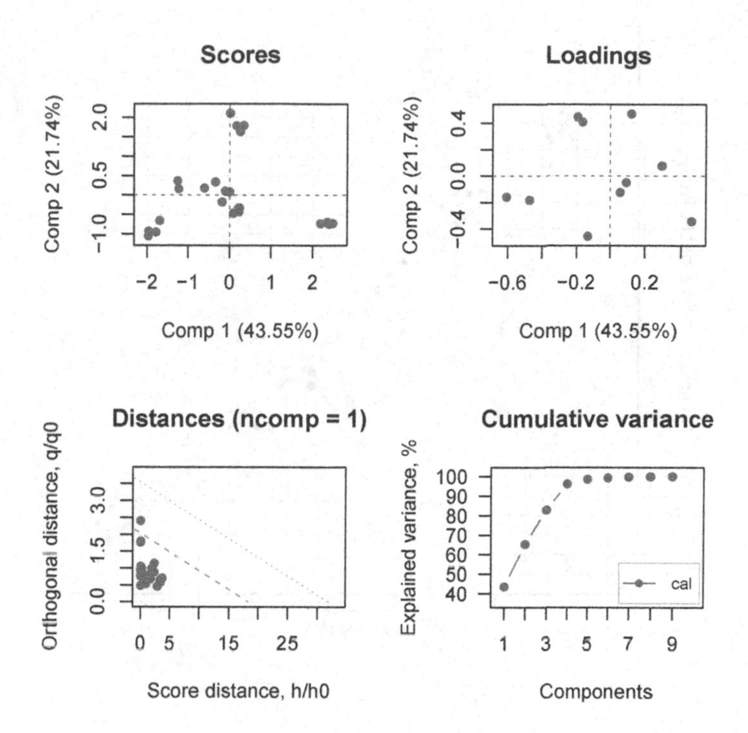

A.5 CHAPTER 4

Linear calibration

```
##
## Call:
## lm(formula = av ~ cs, data = nano.df)
##
## Residuals:
##      Min       1Q  Median      3Q     Max
## -83.271 -16.318   2.307  16.072  45.333
##
## Coefficients:
##              Estimate Std. Error t value Pr(>|t|)
## (Intercept)    764.50      19.36   39.49   <2e-16 ***
## cs           -1023.96      46.19  -22.17   <2e-16 ***
## ---
## Signif. codes:  0 '***' 0.001 '**' 0.01 '*' 0.05 '.' 0.1 ' ' 1
##
## Residual standard error: 26.76 on 46 degrees of freedom
## Multiple R-squared:  0.9144, Adjusted R-squared:  0.9125
## F-statistic: 491.4 on 1 and 46 DF,  p-value: < 2.2e-16
```

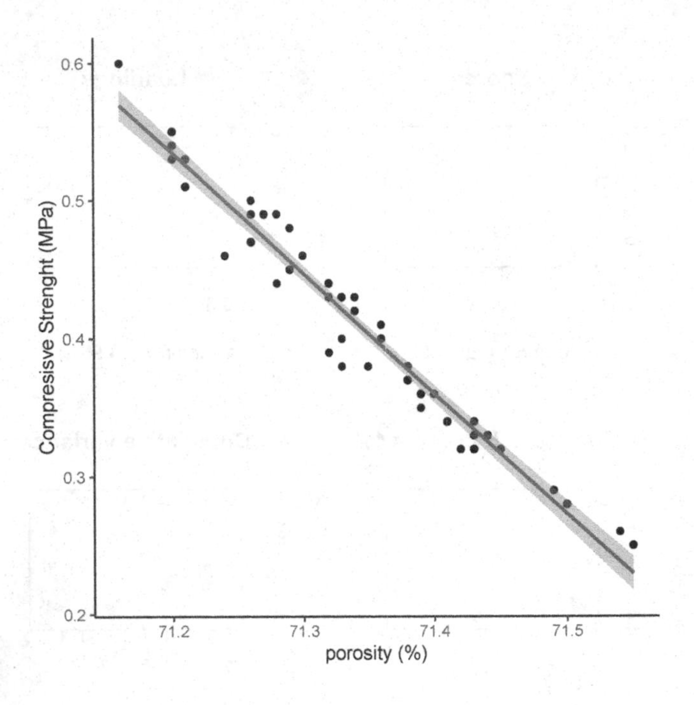


```
##
## Call:
## lm(formula = por ~ cs, data = nano.df)
##
## Residuals:
##       Min        1Q    Median        3Q       Max
## -0.047932 -0.010251  0.004544  0.012525  0.031878
##
## Coefficients:
##             Estimate Std. Error t value Pr(>|t|)
## (Intercept) 71.79437    0.01414 5078.89   <2e-16 ***
## cs          -1.10095    0.03373  -32.64   <2e-16 ***
## ---
## Signif. codes:  0 '***' 0.001 '**' 0.01 '*' 0.05 '.' 0.1 ' ' 1
##
## Residual standard error: 0.01954 on 46 degrees of freedom
## Multiple R-squared:  0.9586, Adjusted R-squared:  0.9577
## F-statistic:  1065 on 1 and 46 DF,  p-value: < 2.2e-16
```

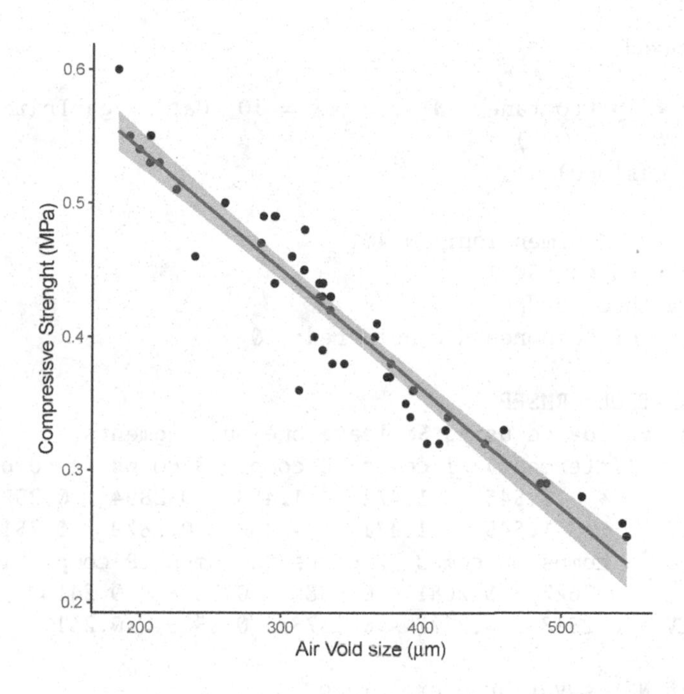


```
##
## Call:
## lm(formula = av ~ cs, data = nano.df)
##
## Residuals:
##      Min       1Q  Median      3Q     Max
## -83.271 -16.318   2.307  16.072  45.333
##
## Coefficients:
##             Estimate Std. Error t value Pr(>|t|)
## (Intercept)   764.50      19.36   39.49   <2e-16 ***
## cs          -1023.96      46.19  -22.17   <2e-16 ***
## ---
## Signif. codes:  0 '***' 0.001 '**' 0.01 '*' 0.05 '.' 0.1 ' ' 1
##
## Residual standard error: 26.76 on 46 degrees of freedom
## Multiple R-squared:  0.9144, Adjusted R-squared:  0.9125
## F-statistic: 491.4 on 1 and 46 DF,  p-value: < 2.2e-16
```

Create test and train set for performing multivariate regression

```
library(pls)
data(gasoline)
gasTrain <- gasoline[1:50,]
gasTest <- gasoline[51:60,]
```

PCR model

```
gas1pcr <- pcr(octane ~ NIR, ncomp = 10, data = gasTrain,
validation = "LOO")
summary(gas1pcr)
```

```
## Data:    X dimension: 50 401
##  Y dimension: 50 1
## Fit method: svdpc
## Number of components considered: 10
##
## VALIDATION: RMSEP
## Cross-validated using 50 leave-one-out segments.
##        (Intercept)  1 comps  2 comps  3 comps  4 comps
## CV           1.545    1.472    1.483   0.2894   0.2522
## adjCV        1.545    1.471    1.482   0.2879   0.2518
##        5 comps  6 comps  7 comps  8 comps  9 comps  10 comps
## CV      0.2622   0.2681   0.2386   0.2328   0.2416    0.2423
## adjCV   0.2618   0.2677   0.2373   0.2323   0.2411    0.2415
##
## TRAINING: % variance explained
##         1 comps  2 comps  3 comps  4 comps  5 comps
## X         79.86    88.12    93.54    96.54    97.74
## octane    16.99    21.36    97.00    97.71    97.73
##         6 comps  7 comps  8 comps  9 comps  10 comps
## X         98.38    98.75    99.06    99.28     99.42
## octane    97.77    98.47    98.54    98.62     98.83
```

Plots PCR models
explained variance

```
plot(RMSEP(gas1pcr), legendpos = "topright")
```

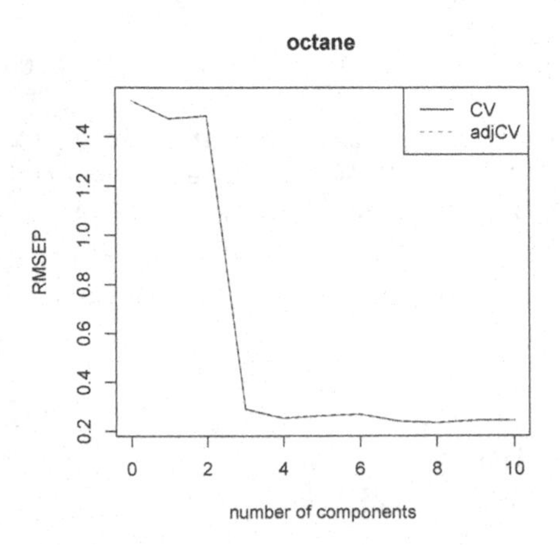

number of components

```
plot(gas1pcr, ncomp = 2, asp = 1, line = TRUE,main="")
```

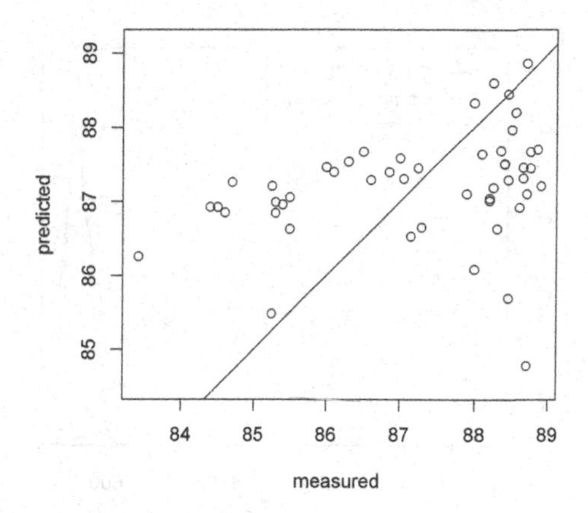

plot of scores

```
plot(gas1pcr, plottype = "scores", comps = 1:3)
```

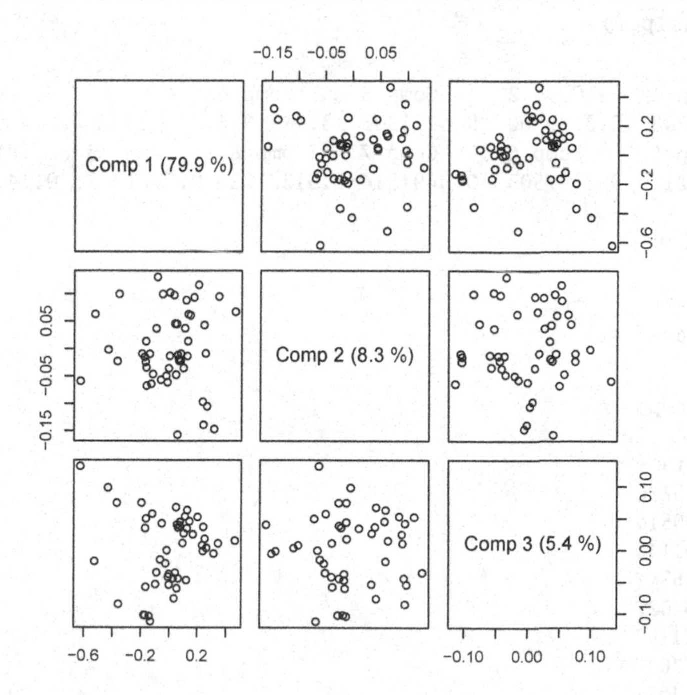

plot of loadings

```
plot(gas1pcr, "loadings", comps = 1:2, legendpos = "topleft",
     labels = "numbers", xlab = "nm")
```

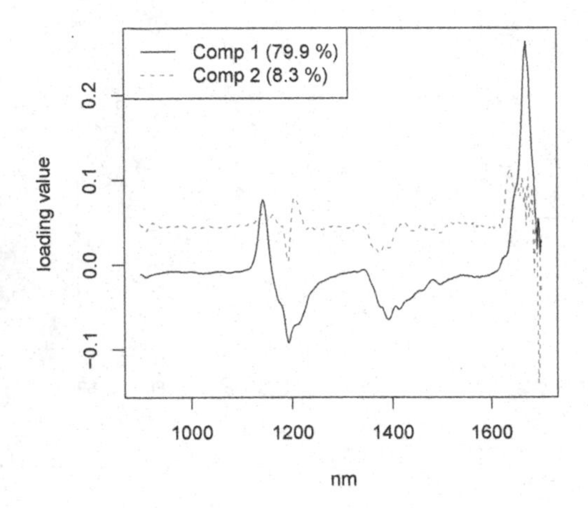

explained variance

```
explvar(gas1pcr)
```

```
##     Comp 1     Comp 2     Comp 3     Comp 4
## 79.8586603  8.2639500  5.4171903  3.0034945
##     Comp 5     Comp 6     Comp 7     Comp 8     Comp 9    Comp 10
##  1.1963215  0.6397503  0.3691514 0.3127762  0.2171267  0.1417888
```

PCR prediction

```
predict(gas1pcr, ncomp = 2,
newdata = gasTest)
```

```
## , , 2 comps
##
##      octane
## 51 87.66776
## 52 87.90510
## 53 87.91140
## 54 86.96307
## 55 87.04627
## 56 86.81198
## 57 87.77647
## 58 87.84051
## 59 88.54958
## 60 87.89335
```

```
predplot(gas1pcr, ncomp = 2, newdata = gasTest, asp = 1,
line = TRUE,main="")
```

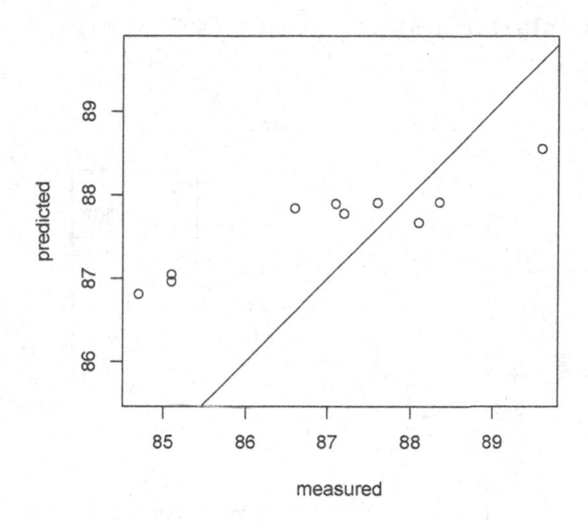

PLS model

```
gas1pls <- plsr(octane ~ NIR, ncomp = 10, data = gasTrain,
validation = "LOO")
summary(gas1pls)
```

```
## Data:     X dimension: 50 401
##  Y dimension: 50 1
## Fit method: kernelpls
## Number of components considered: 10
##
## VALIDATION: RMSEP
## Cross-validated using 50 leave-one-out segments.
##        (Intercept)  1 comps  2 comps  3 comps  4 comps
## CV           1.545    1.357   0.2966   0.2524   0.2476
## adjCV        1.545    1.356   0.2947   0.2521   0.2478


##        5 comps  6 comps 7 comps  8 comps  9 comps  10 comps
## CV      0.2398   0.2319  0.2386   0.2316   0.2449    0.2673
## adjCV.  0.2388   0.2313  0.2377   0.2308   0.2438    0.2657
##
## TRAINING: % variance explained
##         1 comps  2 comps  3 comps  4 comps  5 comps
## X         78.17    85.58    93.41    96.06    96.94
## octane    29.39    96.85    97.89    98.26    98.86
##         6 comps  7 comps  8 comps 9 comps  10 comps
## X         97.89    98.38    98.85   99.02     99.19
## octane    98.96    99.09    99.16   99.28     99.39
```

Plot PLS models
explained variance

```
plot(RMSEP(gas1pls), legendpos = "topright")
```

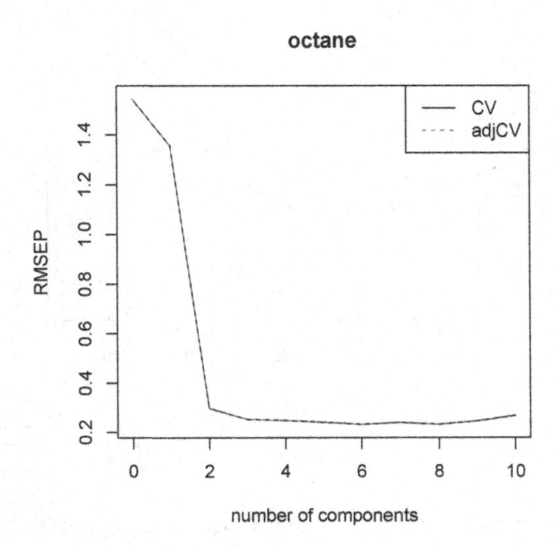

number of components

```
plot(gas1pls, ncomp = 2, asp = 1, line = TRUE,main="")
```

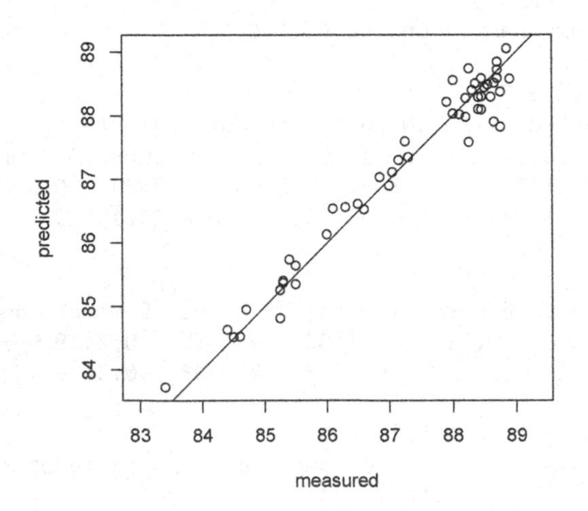

scores plot

```
plot(gas1pls, plottype = "scores", comps = 1:3)
```

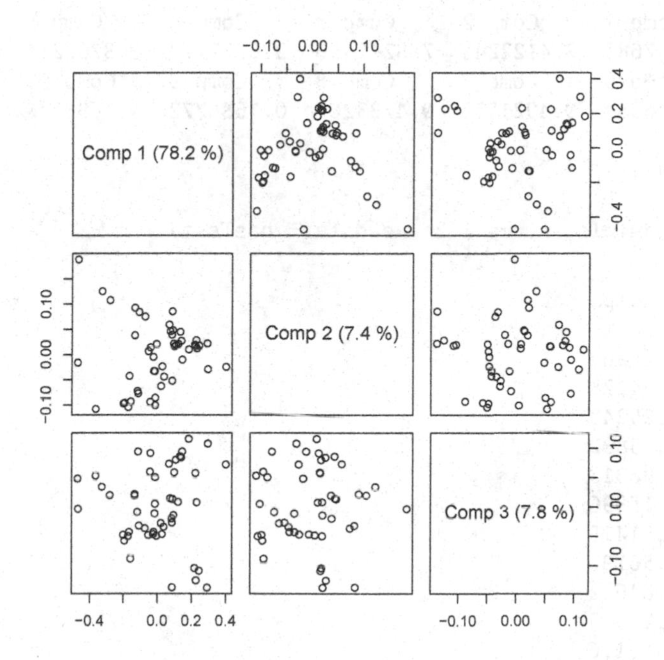

loadings plot

```
plot(gas1pls, "loadings", comps = 1:2,
    legendpos = "topleft",labels = "numbers", xlab = "nm")
```

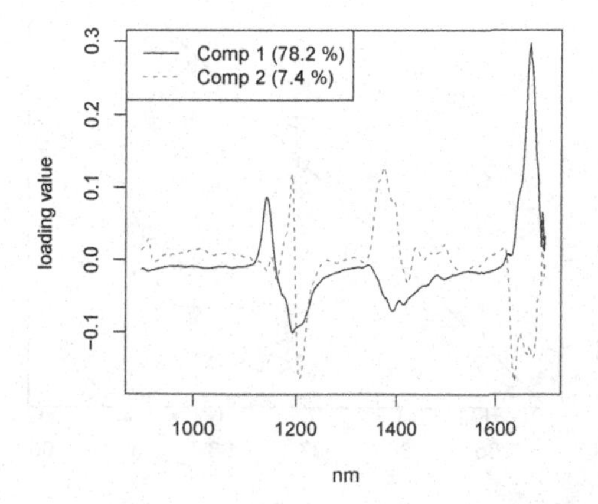

explained variance

```
explvar(gas1pls)
```

```
##      Comp 1     Comp 2     Comp 3     Comp 4     Comp 5
## 78.1707683  7.4122245  7.8241556  2.6577773  0.8768214
##      Comp 6     Comp 7     Comp 8     Comp 9    Comp 10
##  0.9466384  0.4921537  0.4723207  0.1688272  0.1693770
```

pls prediction

```
predict(gas1pls, ncomp = 2, newdata = gasTest)
```

```
## , , 2 comps
##
##      octane
## 51 87.94125
## 52 87.25242
## 53 88.15832
## 54 84.96913
## 55 85.15396
## 56 84.51415
## 57 87.56190
## 58 86.84622
## 59 89.18925
## 60 87.09116
```

```
predplot(gas1pls, ncomp = 2, newdata = gasTest, asp = 1,
line = TRUE,main="")
```

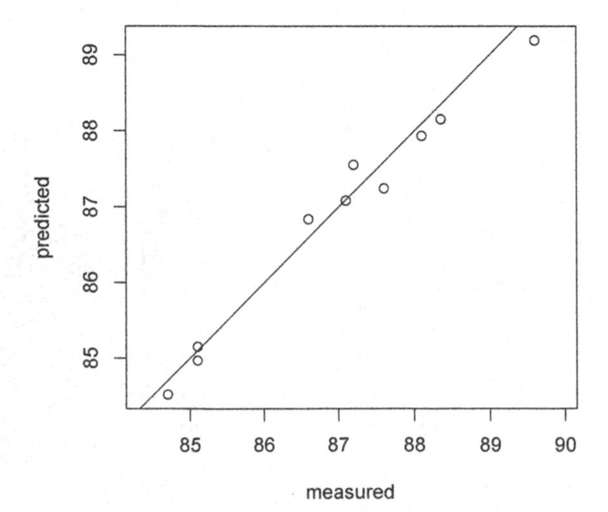

Plot PLS msc models
explained variance

```
plot(RMSEP(gas2), legendpos = "topright")
```

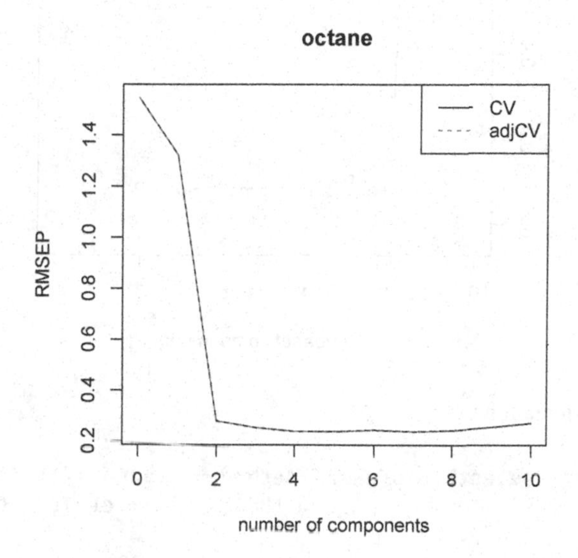

methods of selection for number of components
onesigma

```
ncomp.onesigma <- selectNcomp(gas2,
                  method = "onesigma", plot = TRUE,ylim = c(.18, .6))
```

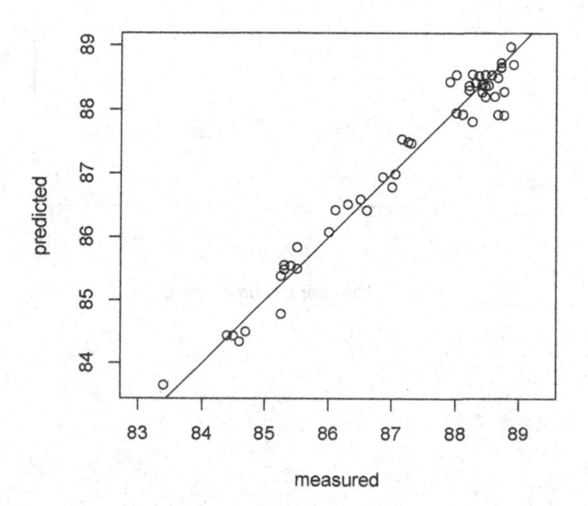

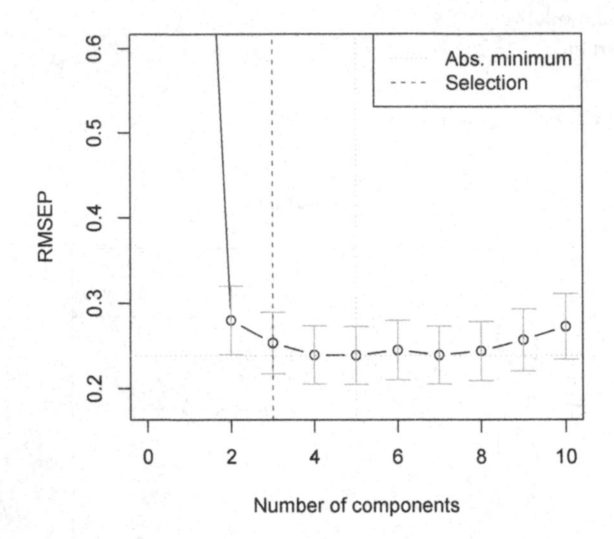

number of permutation

```
ncomp.permut <- selectNcomp(gas2, method = "randomization",
                            plot = TRUE, ylim = c(.18, .6))
```

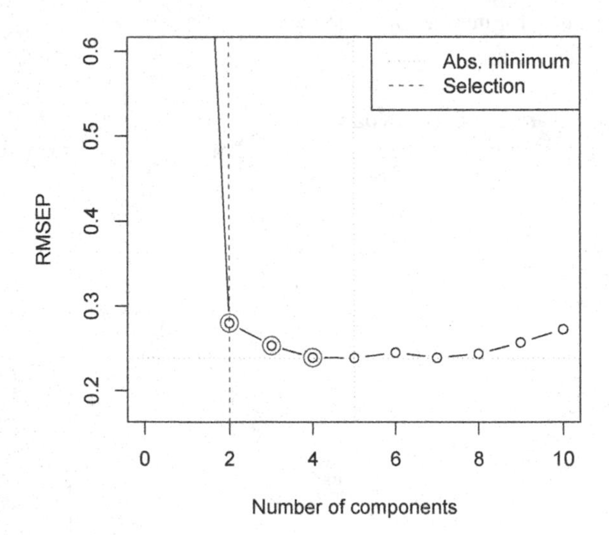

```
gas2.cv <- crossval(gas2, segments = 10)
plot(MSEP(gas2.cv), legendpos="topright")
```

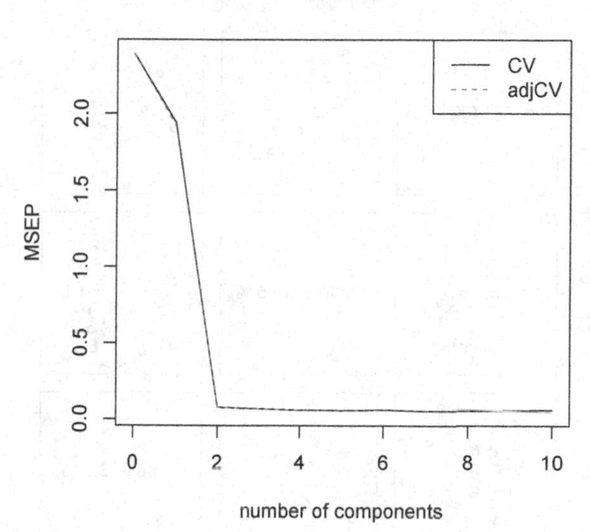


```
summary(gas2.cv, what = "validation")
```

```
## Data:     X dimension: 50 401
##  Y dimension: 50 1
## Fit method: kernelpls
## Number of components considered: 10
##
## VALIDATION: RMSEP
## Cross-validated using 10 random segments.
##         (Intercept)  1 comps  2 comps  3 comps  4 comps  5 comps  6 comps
## CV            1.545    1.337   0.2814   0.2431   0.2384   0.2468   0.2542
## adjCV         1.545    1.333   0.2795   0.2445   0.2355   0.2433   0.2477
##         7 comps  8 comps  9 comps  10 comps
## CV       0.2398   0.2539   0.2711    0.2775
## adjCV    0.2349   0.2474   0.2625    0.2674
```

Scores plot

```
plot(gas2, plottype = "scores", comps = 1:3)
```

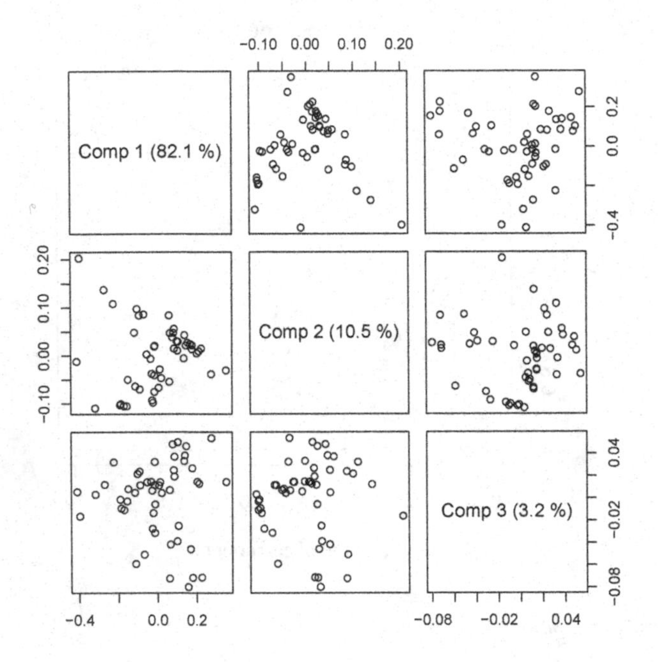

loadings plot

```
plot(gas2, "loadings", comps = 1:2, legendpos = "topleft",xaxt="n",
xlab = "nm")
axis(1, at=c(51,151,251,351), labels=c("1000","1200","1400","1600"))
```

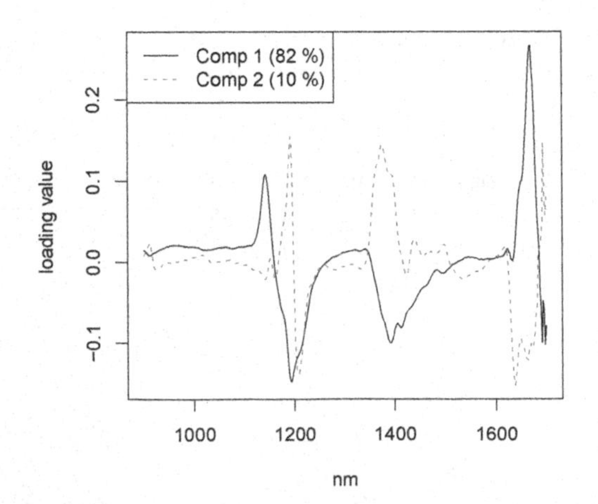

A.6 APPENDIX

Full Factorial n=2 2 levels

```
ff2x2 <- expand.grid(A = c("+", "-"), B = c("+", "-"))
```

Full Factorial n=3 2 levels

```
ff3x2 <- expand.grid(A = c("+", "-"), B = c("+", "-"),
                     C = c("+", "-"))
```

Full Factorial n=4 2 levels

```
ff4x2 <-expand.grid(A = c("+", "-"), B = c("+", "-"),
                    C = c("+", "-"), D = c("+", "-"))
```

Full Factorial n=5 2 levels

```
ff5x2 <- expand.grid(A = c("+", "-"), B = c("+", "-"),
                     C = c("+", "-"),D = c("+", "-"),
                     E= c("+","-"))
```

Fractional Factorial nfactor=4, nruns=8

```
frfr4x8runs <- FrF2(nruns=8,nfactors=4)
```

Fractional Factorial nfactor=5, nruns=16

```
frfr5x16runs <- FrF2(nruns=16,nfactors=5)
```

Fractional Factorial nfactor=6, nruns=32

```
frfr6x32runs <- FrF2(nruns=32,nfactors=6)
```

Fractional Factorial nfactor=7, nruns=64

```
frfr7x16runs <- FrF2(nruns=32,nfactors=7)
```

Plackett-Burman 12

```
placket_burman12 <- pb(12)
```

plackett-burman-16

A.7 PLACKETT-BURMAN 16

```
placket_burman16 <- pb(16)
```

```
## Screening 15 factors in 16 runs involves perfect aliasing of
## pairwise interactions of the first six factors with the last factor.
```

Plackett-Burman 20

```
placket_burman20 <- pb(20)
```

A.8 STATISTICAL TABLES

normal curve areas

```
u <- seq(0,3.09,by=0.01)
p <- pnorm(u)
mp <-matrix(p,ncol=10,byrow=TRUE)
```

critical values for t distribution

```
dft=c(seq(1,100,1))
qt(0.05,dft)
```

```
t005 <- qt(0.05,dft)
t001 <- qt(0.01,dft)
t005 <- qt(0.005,dft)
t001 <- qt(0.001,dft)
```

Critical values for f distribution

```
df1f <- seq(1,100,1)
df2f <- seq(1,100,1)

qf(.95, df1=df1f, df2=df2f)
```

```
f.matrix<- expand.grid("df1"=df1f,"df2"=df2f)

q95 <-qf(.95,df1=f.matrix$df1,df2=f.matrix$df2)
q95m <- matrix(q95,ncol=10,byrow=TRUE)
```

Critical values for chi square

```
chidf <- seq(1,100)
c0995 <- qchisq(0.995,chidf)
```

B A Short Refresher of Matrix Algebra

We will report a minimal refresher of matrix algebra in order to let the readers interested in the details of the methodologies presented in Chapter 3 and 4 better follow the explaination reported.

Examples refer to square matrices due to the fact that *sum-of-squares*, *cross-product*, *variance-covariance*, and *correlation matrices* are all *squares and symmetrical*, and that even while starting with non-square matrices, all calculations will involve their use.

TRACE OF A MATRIX

The *trace of a matrix* is the sum of all the elements (in our case numbers) on the *diagonal starting* from the *upper left to the lowest right value*. If we have the following matrix

$$A = \begin{bmatrix} a_{11} & a_{12} & \cdots \\ \vdots & \ddots & \\ a_{i1} & & a_{ij} \end{bmatrix} = \begin{bmatrix} 1 & 2 & 3 \\ 4 & 5 & 6 \\ 7 & 8 & 9 \end{bmatrix}$$

the trace will be $1+5+9=14$. It is also useful to remark that

- the *trace of the sum-of square matrix* is the *sum of squares*
- the *trace of the cross-product matrix* is the *sum of squares*
- the *trace of the variance-covariance matrix* is the *sum of covariances*
- the *trace of a correlation matrix* is the *number of variables* (each variable contributes a value of 1)

ADDITION OR SUBTRACTION OF A CONSTANT

$$A+k = \begin{bmatrix} a_{11} & a_{12} & \cdots \\ \vdots & \ddots & \\ a_{i1} & & a_{ij} \end{bmatrix} + k = \begin{bmatrix} a_{11}+k & a_{12}+k & \cdots \\ \vdots & \ddots & \\ a_{i1}+k & & a_{ij}+k \end{bmatrix}$$

if $k=3$

$$A = \begin{bmatrix} 1+3 & 2+3 & 3+3 \\ 4+3 & 5+3 & 6+3 \\ 7+3 & 8+3 & 9+3 \end{bmatrix} = \begin{bmatrix} 4 & 5 & 6 \\ 7 & 8 & 9 \\ 10 & 11 & 12 \end{bmatrix}$$

MULTIPLICATION OR DIVISION FOR A CONSTANT

$$A \cdot k = k \cdot \begin{bmatrix} a_{11} & a_{12} & \dots \\ \vdots & \ddots & \\ a_{i1} & & a_{ij} \end{bmatrix} = \begin{bmatrix} k \cdot a_{11} & k \cdot a_{12} & \dots \\ \vdots & \ddots & \\ k \cdot a_{K1} & & k_{ij} \end{bmatrix}$$

if k =3

$$A = \begin{bmatrix} 3 \cdot 1 & 3 \cdot 2 & 3 \cdot 3 \\ 3 \cdot 4 & 3 \cdot 5 & 3 \cdot 6 \\ 3 \cdot 7 & 3 \cdot 8 & 3 \cdot 9 \end{bmatrix} = \begin{bmatrix} 3 & 6 & 9 \\ 12 & 15 & 18 \\ 21 & 24 & 27 \end{bmatrix}$$

ADDITION OR SUBTRACTION OF TWO MATRICES

$$A + B = \begin{bmatrix} a_{11} & a_{12} & \dots \\ \vdots & \ddots & \\ a_{i1} & & a_{ij} \end{bmatrix} + \begin{bmatrix} b_{11} & b_{12} & \dots \\ \vdots & \ddots & \\ b_{i1} & & b_{ij} \end{bmatrix} = \begin{bmatrix} a_{11} + b_{11} & a_{12} + b_{12} & \dots \\ \vdots & \ddots & \\ a_{i1} + b_{i1} & & a_{ij} + b_{ij} \end{bmatrix}$$

$$A + B = \begin{bmatrix} 1 & 2 & 3 \\ 4 & 5 & 6 \\ 7 & 8 & 9 \end{bmatrix} + \begin{bmatrix} 2 & 1 & 0 \\ 4 & 9 & 4 \\ 6 & 7 & 5 \end{bmatrix} = \begin{bmatrix} 3 & 5 & 7 \\ 6 & 14 & 12 \\ 9 & 13 & 14 \end{bmatrix}$$

SCALAR PRODUCT

$$AB = \begin{bmatrix} a_{11} & a_{12} & \dots \\ \vdots & \ddots & \\ a_{i1} & & a_{ij} \end{bmatrix} \begin{bmatrix} b_{11} & b_{12} & \dots \\ \vdots & \ddots & \\ b_{i1} & & b_{ij} \end{bmatrix} =$$

$$= \begin{bmatrix} a_{11} \cdot b_{11} + \dots + a_{1n} \cdot b_{i1} & a_{11} \cdot b_{12} + \dots + a_{1n} \cdot b_{i2} & \dots \\ \vdots & \ddots & \\ a_{i1} \cdot b_{11} + \dots + a_{in} \cdot b_{i1} & & a_{i1} \cdot b_{1j} + \dots + a_{ij} \cdot b_{ij} \end{bmatrix}$$

And with a numerical example:

$$AB = \begin{bmatrix} 1 & 4 & 7 \\ 2 & 5 & 8 \\ 3 & 6 & 9 \end{bmatrix} \begin{bmatrix} 2 & 1 & 0 \\ 4 & 9 & 4 \\ 6 & 7 & 5 \end{bmatrix} =$$

$$= \begin{bmatrix} 1\cdot2+4\cdot4+7\cdot6 & 1\cdot1+4\cdot9+7\cdot7 & 1\cdot0+4\cdot4+7\cdot5 \\ 2\cdot2+5\cdot4+8\cdot6 & 2\cdot1+5\cdot9+8\cdot7 & 2\cdot0+5\cdot4+8\cdot5 \\ 3\cdot2+6\cdot4+9\cdot6 & 3\cdot1+6\cdot9+9\cdot7 & 3\cdot0+6\cdot4+9\cdot5 \end{bmatrix} = \begin{bmatrix} 60 & 86 & 51 \\ 72 & 103 & 60 \\ 84 & 120 & 69 \end{bmatrix}$$

keep in mind that $AB \neq BA$,
and so

$$A = \begin{bmatrix} 2 & 1 & 0 \\ 4 & 9 & 4 \\ 6 & 7 & 5 \end{bmatrix} \begin{bmatrix} 1 & 4 & 7 \\ 2 & 5 & 8 \\ 3 & 6 & 9 \end{bmatrix}$$

$$= \begin{bmatrix} 2\cdot1+1\cdot2+0\cdot3 & 2\cdot4+1\cdot5+0\cdot6 & 2\cdot7+1\cdot8+0\cdot9 \\ 4\cdot1+9\cdot2+4\cdot3 & 4\cdot4+9\cdot5+4\cdot6 & 4\cdot7+9\cdot8+4\cdot9 \\ 6\cdot1+7\cdot2+5\cdot3 & 6\cdot4+7\cdot5+5\cdot6 & 6\cdot7+7\cdot8+5\cdot9 \end{bmatrix} = \begin{bmatrix} 4 & 13 & 22 \\ 34 & 85 & 136 \\ 35 & 89 & 143 \end{bmatrix}$$

MULTIPLICATION OF A MATRIX FOR ITSELF

$$AA = \begin{bmatrix} a_{11} & a_{12} & \cdots \\ \vdots & \ddots & \\ a_{i1} & & a_{ij} \end{bmatrix} \begin{bmatrix} a_{11} & a_{12} & \cdots \\ \vdots & \ddots & \\ a_{i1} & & a_{ij} \end{bmatrix} =$$

$$AA^{\top} = \begin{bmatrix} a_{11}^2 + a_{21} * a12 + \cdots + a_{i1}a1j & a_{11}a_{21} + a_{12}a_{22} + \cdots + a_{1j}a_{i2} & \cdots \\ \vdots & \ddots & \\ a_{i1}a_{11} + a_{i2}a_{21} + \cdots + a_{i1}a_{1j} & \cdots & a_{i1}a_{1j} + a_{i2}a_{2j} + \cdots + a_{ij}^2 \end{bmatrix}$$

Please pay attention while using your software of choice. There are distinct operators for scalar and cross products.

TRANSPOSE MATRIX

$$A = \begin{bmatrix} a_{11} & a_{12} & \cdots \\ \vdots & \ddots & \\ a_{i1} & & a_{ij} \end{bmatrix} = \begin{bmatrix} 1 & 2 & 3 \\ 4 & 5 & 6 \\ 7 & 8 & 9 \end{bmatrix}$$

$$A^{\top} = \begin{bmatrix} a_{11} & a_{21} & \cdots \\ \vdots & \ddots & \\ a_{1j} & & a_{ji} \end{bmatrix} = \begin{bmatrix} 1 & 4 & 7 \\ 2 & 5 & 8 \\ 3 & 6 & 9 \end{bmatrix}$$

MULTIPLICATION OF A MATRIX FOR ITS TRANSPOSED

$$AA^{\intercal} = \begin{bmatrix} a_{11} & a_{21} & \dots \\ \vdots & \ddots & \\ a_{1j} & & a_{ji} \end{bmatrix} \begin{bmatrix} a_{11} & a_{12} & \dots \\ \vdots & \ddots & \\ a_{i1} & & a_{ij} \end{bmatrix}$$

$$AA^{\intercal} = \begin{bmatrix} a_{11}^2 + a_{21}^2 + \cdots + a_{1j}^2 & a_{11}a_{21} + a_{21}a_{22} + \cdots + a_{1j}a_{2i} & \dots \\ \vdots & \ddots & \\ a_{11}a_{1j} + a_{12}a_{2j} + \cdots + a_{1j}a_{ij} & \dots & a_{1j}^2 + a_{2j}^2 + \cdots + a_{ij}^2 \end{bmatrix}$$

INVERSE MATRICES AND DETERMINANTS

We define the identity matrix as a matrix where the diagonal is made up of 1s and all other values are 0

$$I = \begin{bmatrix} 1 & 0 & \dots \\ \vdots & \ddots & \\ 0 & & 1 \end{bmatrix}$$

In order to find the inverse of a matrix we need to satisfy the following condition

$$A^{-1}A = AA^{-1} = I$$

One method for finding a solution involves the calculation of the *determinant* of A, noted as $|A|$. For a matrix of dimension 2x2, we can write that

$$A = \begin{bmatrix} a_{11} & a_{12} \\ a_{i1} & a_{ij} \end{bmatrix}$$

$$|A| = a_{11}a_{22} - a_{12}a_{21}$$

For calculating the determinant of a 3x3 matrix we can use the following formula.

$$A = \begin{bmatrix} a_{11} & a_{12} & a_{13} \\ a_{21} & a_{22} & a_{23} \\ a_{31} & a_{42} & a_{33} \end{bmatrix} = a_{11}(a_{22}a_{33} - a_{23}a_{32}) - a_{12}(a_{21}a_{33} - a_{23}a_{31}) + a_{13}(a_{21}a_{32} - a_{23}a_{31})$$

Due to the properties of determinant we can also write

$$A^{-1} = \begin{bmatrix} a_{11} & a_{12} & a_{13} \\ a_{21} & a_{22} & a_{23} \\ a_{31} & a_{42} & a_{33} \end{bmatrix}^{-1} = \frac{1}{|A|} \begin{bmatrix} a_{22}a_{33} - a_{23}a_{32} & a_{13}a_{32} - a_{12}a_{33} & a_{12}a_{23} - a_{13}a_{22} \\ a_{23}a_{11} - a_{12}a_{33} & a_{11}a_{33} - a_{13}a_{31} & a_{13}a_{21} - a_{11}a_{23} \\ a_{21}a_{32} - a_{22}a_{31} & a_{12}a_{31} - a_{11}a_{31} & a_{11}a_{22} - a_{12}a_{21} \end{bmatrix}$$

in numbers, considering that

$$|A| = 2(4\cdot8-3\cdot7)+5(7\cdot4-3\cdot8)+9(3\cdot3-4\cdot4) = (22+20-63) = -21$$

$$A^{-1} = \begin{bmatrix} 2 & 5 & 9 \\ 3 & 4 & 7 \\ 4 & 3 & 8 \end{bmatrix}^{-1} = \frac{1}{-21}\begin{bmatrix} 4\cdot8-7\cdot3 & 9\cdot3-5\cdot8 & 5\cdot7-9\cdot4 \\ 7\cdot4-8\cdot3 & 2\cdot8-9\cdot4 & 9\cdot3-2\cdot7 \\ 3\cdot3-4\cdot4 & 5\cdot4-2\cdot3 & 2\cdot4-5\cdot3 \end{bmatrix}$$

$$\frac{1}{-21}\begin{bmatrix} 11 & -13 & -1 \\ 4 & -20 & 13 \\ -7 & 14 & -7 \end{bmatrix} = \begin{bmatrix} -0.523 & 0.619 & 0.048 \\ -0.195 & 0.952 & -0.619 \\ 0.333 & -0.667 & -0.333 \end{bmatrix}$$

What useful information can give us the value of the *determinant* about inverting a matrix?

If it is equal to 0, we cannot invert our matrix, since its inversion will involve division by 0.

EIGENVALUES

In order to find the solution of an eigenproblem, we need to solve the following equation

$$(D-\lambda I)V = 0$$

where λ is the eigenvalue and V is the eigenvector to be found. Expanding the equation we have:

$$\left[\begin{bmatrix} a_{11} & a_{12} \\ a_{i1} & a_{ij} \end{bmatrix} - \lambda \begin{bmatrix} 1 & 0 \\ 0 & 1 \end{bmatrix}\right]\begin{bmatrix} v_1 \\ v_2 \end{bmatrix} = 0$$

or

$$\left[\begin{bmatrix} a_{11} & a_{12} \\ a_{i1} & a_{ij} \end{bmatrix} - \begin{bmatrix} \lambda & 0 \\ 0 & \lambda \end{bmatrix}\right]\begin{bmatrix} v_1 \\ v_2 \end{bmatrix} = 0$$

$$\begin{bmatrix} a_{11}-\lambda & a_{12} \\ a_{i1} & a_{ij}- \end{bmatrix}\begin{bmatrix} v_1 \\ v_2 \end{bmatrix} = 0$$

$$(a_{11}-\lambda a_{ij}-) - a_{12}a_{i1} = 0 \tag{B.1}$$

$$\lambda^2 - (a_{11}+a_{ij})\lambda) + a_{11}a_{ij} - a_{12}a_{i1} = 0 \tag{B.2}$$

that can be written as

$$x^2 + y\lambda + z = 0 \tag{B.3}$$

In order to find the root we can use

$$\lambda = \frac{-y \pm \sqrt{y^2 - 4xz}}{2x} \tag{B.4}$$

a numerical example

$$D = \begin{bmatrix} 2 & 3 \\ 5 & 4 \end{bmatrix}$$

Applying previously seen equation

$$\lambda^2 - (2+4)\lambda + 2 \cdot 4 - 3 \cdot 5 = 0 \tag{B.5}$$

or

$$\lambda^2 - 6\lambda - 7$$

$$\lambda = \frac{-(-6) + \sqrt{((-6)^2 - 4 \cdot (-7) \cdot 1}}{2 \cdot 1} = 7 \tag{B.6}$$

$$\lambda = \frac{-(-6) - \sqrt{((-6)^2 - 4 \cdot (-7) \cdot 1}}{2 \cdot 1} = -1 \tag{B.7}$$

$$-5v1 + 3v2$$

$$5v1 - 3v2$$

Substituting the first root -1 found

$$\begin{bmatrix} 2-(-1) & 3 \\ 5 & 4-(-1) \end{bmatrix} \begin{bmatrix} v_1 \\ v_2 \end{bmatrix} = 0$$

$$\begin{bmatrix} -5 & 3 \\ 5 & -3 \end{bmatrix} \begin{bmatrix} v_1 \\ v_2 \end{bmatrix}$$

$$-5v1 + 3v2$$

$$5v1 - 3v2$$

The first solution is $x = 3n$ and $y = 5nn \in \mathbb{Z}$

Analogously it is possible to substitute the second root and find the second eigenvectors. This example was for a 2×2 matrix, and consequently, the polynomial for eigenvalues was quadratic, and there were two equations with two unknowns to solve the equation. As the dimensions of the matrix increase, the number, degree, and unknowns also increase.

For a visual approach regarding the properties of eigenvalues and eigenvector, I suggest to visit the website by Victor Powell and Lewis Lehe at the address setosa.io/ev/eigenvectors-and-eigenvalues/, where you will also find interactive visualizations that can help to intuitively grasp the concepts presented here.

C Statistical Tables

Table C.1
Normal curve areas.

	0	0.01	0.02	0.03	0.04	0.05	0.06	0.07	0.08	0.09
-3.40	0.0003	0.0003	0.0003	0.0003	0.0003	0.0003	0.0003	0.0003	0.0003	0.0002
-3.30	0.0005	0.0005	0.0005	0.0004	0.0004	0.0004	0.0004	0.0004	0.0004	0.0003
-3.20	0.0007	0.0007	0.0006	0.0006	0.0006	0.0006	0.0006	0.0005	0.0005	0.0005
-3.10	0.0010	0.0009	0.0009	0.0009	0.0008	0.0008	0.0008	0.0008	0.0007	0.0007
-3.00	0.0013	0.0013	0.0013	0.0012	0.0012	0.0011	0.0011	0.0011	0.0010	0.0010
-2.90	0.0019	0.0018	0.0018	0.0017	0.0016	0.0016	0.0015	0.0015	0.0014	0.0014
-2.80	0.0026	0.0025	0.0024	0.0023	0.0023	0.0022	0.0021	0.0021	0.0020	0.0019
-2.70	0.0035	0.0034	0.0033	0.0032	0.0031	0.0030	0.0029	0.0028	0.0027	0.0026
-2.60	0.0047	0.0045	0.0044	0.0043	0.0041	0.0040	0.0039	0.0038	0.0037	0.0036
-2.50	0.0062	0.0060	0.0059	0.0057	0.0055	0.0054	0.0052	0.0051	0.0049	0.0048
-2.40	0.0082	0.0080	0.0078	0.0075	0.0073	0.0071	0.0069	0.0068	0.0066	0.0064
-2.30	0.0107	0.0104	0.0102	0.0099	0.0096	0.0094	0.0091	0.0089	0.0087	0.0084
-2.20	0.0139	0.0136	0.0132	0.0129	0.0125	0.0122	0.0119	0.0116	0.0113	0.0110
-2.10	0.0179	0.0174	0.0170	0.0166	0.0162	0.0158	0.0154	0.0150	0.0146	0.0143
-2.00	0.0228	0.0222	0.0217	0.0212	0.0207	0.0202	0.0197	0.0192	0.0188	0.0183
-1.90	0.0287	0.0281	0.0274	0.0268	0.0262	0.0256	0.0250	0.0244	0.0239	0.0233
-1.80	0.0359	0.0351	0.0344	0.0336	0.0329	0.0322	0.0314	0.0307	0.0301	0.0294
-1.70	0.0446	0.0436	0.0427	0.0418	0.0409	0.0401	0.0392	0.0384	0.0375	0.0367
-1.60	0.0548	0.0537	0.0526	0.0516	0.0505	0.0495	0.0485	0.0475	0.0465	0.0455
-1.50	0.0668	0.0655	0.0643	0.0630	0.0618	0.0606	0.0594	0.0582	0.0571	0.0559
-1.40	0.0808	0.0793	0.0778	0.0764	0.0749	0.0735	0.0721	0.0708	0.0694	0.0681
-1.30	0.0968	0.0951	0.0934	0.0918	0.0901	0.0885	0.0869	0.0853	0.0838	0.0823
-1.20	0.1151	0.1131	0.1112	0.1093	0.1075	0.1056	0.1038	0.1020	0.1003	0.0985
-1.10	0.1357	0.1335	0.1314	0.1292	0.1271	0.1251	0.1230	0.1210	0.1190	0.1170
-1.00	0.1587	0.1562	0.1539	0.1515	0.1492	0.1469	0.1446	0.1423	0.1401	0.1379
-0.90	0.1841	0.1814	0.1788	0.1762	0.1736	0.1711	0.1685	0.1660	0.1635	0.1611
-0.80	0.2119	0.2090	0.2061	0.2033	0.2005	0.1977	0.1949	0.1922	0.1894	0.1867
-0.70	0.2420	0.2389	0.2358	0.2327	0.2296	0.2266	0.2236	0.2206	0.2177	0.2148
-0.60	0.2743	0.2709	0.2676	0.2643	0.2611	0.2578	0.2546	0.2514	0.2483	0.2451
-0.50	0.3085	0.3050	0.3015	0.2981	0.2946	0.2912	0.2877	0.2843	0.2810	0.2776
-0.40	0.3446	0.3409	0.3372	0.3336	0.3300	0.3264	0.3228	0.3192	0.3156	0.3121
-0.30	0.3821	0.3783	0.3745	0.3707	0.3669	0.3632	0.3594	0.3557	0.3520	0.3483
-0.20	0.4207	0.4168	0.4129	0.4090	0.4052	0.4013	0.3974	0.3936	0.3897	0.3859
-0.10	0.4602	0.4562	0.4522	0.4483	0.4443	0.4404	0.4364	0.4325	0.4286	0.4247
0.00	0.5000	0.4960	0.4920	0.4880	0.4840	0.4801	0.4761	0.4721	0.4681	0.4641

Table C.2
Critical values of the t distribution for α = 0.05, 0.01, 0.005, 0.001, lower tail $P[X \leq x]$.

dof	0.05	0.01	0.005	0.001
1	-6.313752	-31.820516	-63.656741	-318.308839
2	-2.919986	-6.964557	-9.924843	-22.327125
3	-2.353363	-4.540703	-5.840909	-10.214532
4	-2.131847	-3.746947	-4.604095	-7.173182
5	-2.015048	-3.364930	-4.032143	-5.893429
6	-1.943180	-3.142668	-3.707428	-5.207626
7	-1.894579	-2.997952	-3.499483	-4.785290
8	-1.859548	-2.896459	-3.355387	-4.500791
9	-1.833113	-2.821438	-3.249836	-4.296806
10	-1.812461	-2.763769	-3.169273	-4.143700
11	-1.795885	-2.718079	-3.105806	-4.024701
12	-1.782288	-2.680998	-3.054540	-3.929633
13	-1.770933	-2.650309	-3.012276	-3.851982
14	-1.761310	-2.624494	-2.976843	-3.787390
15	-1.753050	-2.602480	-2.946713	-3.732834
16	-1.745884	-2.583487	-2.920782	-3.686155
17	-1.739607	-2.566934	-2.898230	-3.645767
18	-1.734064	-2.552380	-2.878440	-3.610485
19	-1.729133	-2.539483	-2.860935	-3.579400
20	-1.724718	-2.527977	-2.845340	-3.551808
21	-1.720743	-2.517648	-2.831360	-3.527154
22	-1.717144	-2.508325	-2.818756	-3.504992
23	-1.713871	-2.499867	-2.807336	-3.484964
24	-1.710882	-2.492160	-2.796940	-3.466777
25	-1.708141	-2.485107	-2.787436	-3.450189
26	-1.705618	-2.478630	-2.778714	-3.434997
27	-1.703288	-2.472660	-2.770683	-3.421034
28	-1.701131	-2.467140	-2.763263	-3.408155
29	-1.699127	-2.462021	-2.756386	-3.396240
30	-1.697261	-2.457261	-2.749996	-3.385185
30	-1.697261	-2.457261	-2.749996	-3.385185
40	-1.683851	-2.423257	-2.704459	-3.306878
60	-1.670649	-2.390119	-2.660283	-3.231709
120	-1.657651	-2.357825	-2.617421	-3.159539

Table C.3
Critical values of the F distribution p = 0.95.

	1	2	3	4	5	6	7	12	24	120
1	161.45	199.50	215.71	224.58	230.16	233.99	236.77	243.91	249.05	253.25
2	18.51	19.00	19.16	19.25	19.30	19.33	19.35	19.41	19.45	19.49
3	10.13	9.55	9.28	9.12	9.01	8.94	8.89	8.74	8.64	8.55
4	7.71	6.94	6.59	6.39	6.26	6.16	6.09	5.91	5.77	5.66
5	6.61	5.79	5.41	5.19	5.05	4.95	4.88	4.68	4.53	4.40
6	5.99	5.14	4.76	4.53	4.39	4.28	4.21	4.00	3.84	3.70
7	5.59	4.74	4.35	4.12	3.97	3.87	3.79	3.57	3.41	3.27
8	5.32	4.46	4.07	3.84	3.69	3.58	3.50	3.28	3.12	2.97
9	5.12	4.26	3.86	3.63	3.48	3.37	3.29	3.07	2.90	2.75
10	4.96	4.10	3.71	3.48	3.33	3.22	3.14	2.91	2.74	2.58
11	4.84	3.98	3.59	3.36	3.20	3.09	3.01	2.79	2.61	2.45
12	4.75	3.89	3.49	3.26	3.11	3.00	2.91	2.69	2.51	2.34
13	4.67	3.81	3.41	3.18	3.03	2.92	2.83	2.60	2.42	2.25
14	4.60	3.74	3.34	3.11	2.96	2.85	2.76	2.53	2.35	2.18
15	4.54	3.68	3.29	3.06	2.90	2.79	2.71	2.48	2.29	2.11
16	4.49	3.63	3.24	3.01	2.85	2.74	2.66	2.42	2.24	2.06
17	4.45	3.59	3.20	2.96	2.81	2.70	2.61	2.38	2.19	2.01
18	4.41	3.55	3.16	2.93	2.77	2.66	2.58	2.34	2.15	1.97
19	4.38	3.52	3.13	2.90	2.74	2.63	2.54	2.31	2.11	1.93
20	4.35	3.49	3.10	2.87	2.71	2.60	2.51	2.28	2.08	1.90
21	4.32	3.47	3.07	2.84	2.68	2.57	2.49	2.25	2.05	1.87
22	4.30	3.44	3.05	2.82	2.66	2.55	2.46	2.23	2.03	1.84
23	4.28	3.42	3.03	2.80	2.64	2.53	2.44	2.20	2.01	1.81
24	4.26	3.40	3.01	2.78	2.62	2.51	2.42	2.18	1.98	1.79
25	4.24	3.39	2.99	2.76	2.60	2.49	2.40	2.16	1.96	1.77
26	4.23	3.37	2.98	2.74	2.59	2.47	2.39	2.15	1.95	1.75
27	4.21	3.35	2.96	2.73	2.57	2.46	2.37	2.13	1.93	1.73
28	4.20	3.34	2.95	2.71	2.56	2.45	2.36	2.12	1.91	1.71
29	4.18	3.33	2.93	2.70	2.55	2.43	2.35	2.10	1.90	1.70
30	4.17	3.32	2.92	2.69	2.53	2.42	2.33	2.09	1.89	1.68
40	4.08	3.23	2.84	2.61	2.45	2.34	2.25	2.00	1.79	1.58
60	4.00	3.15	2.76	2.53	2.37	2.25	2.17	1.92	1.70	1.47
120	3.92	3.07	2.68	2.45	2.29	2.18	2.09	1.83	1.61	1.35

Table C.4

Critical values of the F distribution p = 0.99, with rows and columns reports degree of freedom.

	1	2	3	4	5	6	7	12	24	120
1	4052.2	4999.0	5403.3	5624.6	5763.6	5859.0	5928.4	6106.3	6234.6	6339.4
2	98.50	99.00	99.17	99.25	99.30	99.33	99.36	99.42	99.46	99.49
3	34.12	30.82	29.46	28.71	28.24	27.91	27.67	27.05	26.60	26.22
4	21.20	18.00	16.69	15.98	15.52	15.21	14.98	14.37	13.93	13.56
5	16.26	13.27	12.06	11.39	10.97	10.67	10.46	9.89	9.47	9.11
6	13.75	10.92	9.78	9.15	8.75	8.47	8.26	7.72	7.31	6.97
7	12.25	9.55	8.45	7.85	7.46	7.19	6.99	6.47	6.07	5.74
8	11.26	8.65	7.59	7.01	6.63	6.37	6.18	5.67	5.28	4.95
9	10.56	8.02	6.99	6.42	6.06	5.80	5.61	5.11	4.73	4.40
10	10.04	7.56	6.55	5.99	5.64	5.39	5.20	4.71	4.33	4.00
11	9.65	7.21	6.22	5.67	5.32	5.07	4.89	4.40	4.02	3.69
12	9.33	6.93	5.95	5.41	5.06	4.82	4.64	4.16	3.78	3.45
13	9.07	6.70	5.74	5.21	4.86	4.62	4.44	3.96	3.59	3.25
14	8.86	6.51	5.56	5.04	4.69	4.46	4.28	3.80	3.43	3.09
15	8.68	6.36	5.42	4.89	4.56	4.32	4.14	3.67	3.29	2.96
16	8.53	6.23	5.29	4.77	4.44	4.20	4.03	3.55	3.18	2.84
17	8.40	6.11	5.18	4.67	4.34	4.10	3.93	3.46	3.08	2.75
18	8.29	6.01	5.09	4.58	4.25	4.01	3.84	3.37	3.00	2.66
19	8.18	5.93	5.01	4.50	4.17	3.94	3.77	3.30	2.92	2.58
20	8.10	5.85	4.94	4.43	4.10	3.87	3.70	3.23	2.86	2.52
21	8.02	5.78	4.87	4.37	4.04	3.81	3.64	3.17	2.80	2.46
22	7.95	5.72	4.82	4.31	3.99	3.76	3.59	3.12	2.75	2.40
23	7.88	5.66	4.76	4.26	3.94	3.71	3.54	3.07	2.70	2.35
24	7.82	5.61	4.72	4.22	3.90	3.67	3.50	3.03	2.66	2.31
25	7.77	5.57	4.68	4.18	3.85	3.63	3.46	2.99	2.62	2.27
26	7.72	5.53	4.64	4.14	3.82	3.59	3.42	2.96	2.58	2.23
27	7.68	5.49	4.60	4.11	3.78	3.56	3.39	2.93	2.55	2.20
28	7.64	5.45	4.57	4.07	3.75	3.53	3.36	2.90	2.52	2.17
29	7.60	5.42	4.54	4.04	3.73	3.50	3.33	2.87	2.49	2.14
30	7.56	5.39	4.51	4.02	3.70	3.47	3.30	2.84	2.47	2.11
40	7.31	5.18	4.31	3.83	3.51	3.29	3.12	2.66	2.29	1.92
60	7.08	4.98	4.13	3.65	3.34	3.12	2.95	2.50	2.12	1.73
120	6.85	4.79	3.95	3.48	3.17	2.96	2.79	2.34	1.95	1.53

Table C.5

Critical values of the F distribution p = 0.995, with rows and columns reports degree of freedom.

	1	2	3	4	5	6	7	12	24	120
1	1.62e4	1.99e4	2.16e4	2.24e4	2.30e4	2.34e4	2.37e4	2.44e4	2.49e4	2.53e4
2	198.50	199.00	199.17	199.25	199.30	199.33	199.36	199.42	199.46	199.49
3	55.55	49.80	47.47	46.19	45.39	44.84	44.43	43.39	42.62	41.99
4	31.33	26.28	24.26	23.15	22.46	21.97	21.62	20.70	20.03	19.47
5	22.78	18.31	16.53	15.56	14.94	14.51	14.20	13.38	12.78	12.27
6	18.63	14.54	12.92	12.03	11.46	11.07	10.79	10.03	9.47	9.00
7	16.24	12.40	10.88	10.05	9.52	9.16	8.89	8.18	7.64	7.19
8	14.69	11.04	9.60	8.81	8.30	7.95	7.69	7.01	6.50	6.06
9	13.61	10.11	8.72	7.96	7.47	7.13	6.88	6.23	5.73	5.30
10	12.83	9.43	8.08	7.34	6.87	6.54	6.30	5.66	5.17	4.75
11	12.23	8.91	7.60	6.88	6.42	6.10	5.86	5.24	4.76	4.34
12	11.75	8.51	7.23	6.52	6.07	5.76	5.52	4.91	4.43	4.01
13	11.37	8.19	6.93	6.23	5.79	5.48	5.25	4.64	4.17	3.76
14	11.06	7.92	6.68	6.00	5.56	5.26	5.03	4.43	3.96	3.55
15	10.80	7.70	6.48	5.80	5.37	5.07	4.85	4.25	3.79	3.37
16	10.58	7.51	6.30	5.64	5.21	4.91	4.69	4.10	3.64	3.22
17	10.38	7.35	6.16	5.50	5.07	4.78	4.56	3.97	3.51	3.10
18	10.22	7.21	6.03	5.37	4.96	4.66	4.44	3.86	3.40	2.99
19	10.07	7.09	5.92	5.27	4.85	4.56	4.34	3.76	3.31	2.89
20	9.94	6.99	5.82	5.17	4.76	4.47	4.26	3.68	3.22	2.81
21	9.83	6.89	5.73	5.09	4.68	4.39	4.18	3.60	3.15	2.73
22	9.73	6.81	5.65	5.02	4.61	4.32	4.11	3.54	3.08	2.66
23	9.63	6.73	5.58	4.95	4.54	4.26	4.05	3.47	3.02	2.60
24	9.55	6.66	5.52	4.89	4.49	4.20	3.99	3.42	2.97	2.55
25	9.48	6.60	5.46	4.84	4.43	4.15	3.94	3.37	2.92	2.50
26	9.41	6.54	5.41	4.79	4.38	4.10	3.89	3.33	2.87	2.45
27	9.34	6.49	5.36	4.74	4.34	4.06	3.85	3.28	2.83	2.41
28	9.28	6.44	5.32	4.70	4.30	4.02	3.81	3.25	2.79	2.37
29	9.23	6.40	5.28	4.66	4.26	3.98	3.77	3.21	2.76	2.33
30	9.18	6.35	5.24	4.62	4.23	3.95	3.74	3.18	2.73	2.30
40	8.83	6.07	4.98	4.37	3.99	3.71	3.51	2.95	2.50	2.06
60	8.49	5.79	4.73	4.14	3.76	3.49	3.29	2.74	2.29	1.83
120	8.18	5.54	4.50	3.92	3.55	3.28	3.09	2.54	2.09	1.61

Table C.6

Critical values of the F distribution p = 0.999, with rows and columns reports degree of freedom.

	1	2	3	4	5	6	7	12	24	120
1	4.05e5	4.99e5	5.40e5	5.62e5	5.76e5	5.85e5	5.92e5	6.10e5	6.23e5	6.33e5
2	998.50	999.00	999.17	999.25	999.30	999.33	999.36	999.42	999.46	999.49
3	167.03	148.50	141.11	137.10	134.58	132.85	131.58	128.32	125.93	123.97
4	74.14	61.25	56.18	53.44	51.71	50.53	49.66	47.41	45.77	44.40
5	47.18	37.12	33.20	31.09	29.75	28.83	28.16	26.42	25.13	24.06
6	35.51	27.00	23.70	21.92	20.80	20.03	19.46	17.99	16.90	15.98
7	29.25	21.69	18.77	17.20	16.21	15.52	15.02	13.71	12.73	11.91
8	25.41	18.49	15.83	14.39	13.48	12.86	12.40	11.19	10.30	9.53
9	22.86	16.39	13.90	12.56	11.71	11.13	10.70	9.57	8.72	8.00
10	21.04	14.91	12.55	11.28	10.48	9.93	9.52	8.45	7.64	6.94
11	19.69	13.81	11.56	10.35	9.58	9.05	8.66	7.63	6.85	6.18
12	18.64	12.97	10.80	9.63	8.89	8.38	8.00	7.00	6.25	5.59
13	17.82	12.31	10.21	9.07	8.35	7.86	7.49	6.52	5.78	5.14
14	17.14	11.78	9.73	8.62	7.92	7.44	7.08	6.13	5.41	4.77
15	16.59	11.34	9.34	8.25	7.57	7.09	6.74	5.81	5.10	4.47
16	16.12	10.97	9.01	7.94	7.27	6.80	6.46	5.55	4.85	4.23
17	15.72	10.66	8.73	7.68	7.02	6.56	6.22	5.32	4.63	4.02
18	15.38	10.39	8.49	7.46	6.81	6.35	6.02	5.13	4.45	3.84
19	15.08	10.16	8.28	7.27	6.62	6.18	5.85	4.97	4.29	3.68
20	14.82	9.95	8.10	7.10	6.46	6.02	5.69	4.82	4.15	3.54
21	14.59	9.77	7.94	6.95	6.32	5.88	5.56	4.70	4.03	3.42
22	14.38	9.61	7.80	6.81	6.19	5.76	5.44	4.58	3.92	3.32
23	14.20	9.47	7.67	6.70	6.08	5.65	5.33	4.48	3.82	3.22
24	14.03	9.34	7.55	6.59	5.98	5.55	5.23	4.39	3.74	3.14
25	13.88	9.22	7.45	6.49	5.89	5.46	5.15	4.31	3.66	3.06
26	13.74	9.12	7.36	6.41	5.80	5.38	5.07	4.24	3.59	2.99
27	13.61	9.02	7.27	6.33	5.73	5.31	5.00	4.17	3.52	2.92
28	13.50	8.93	7.19	6.25	5.66	5.24	4.93	4.11	3.46	2.86
29	13.39	8.85	7.12	6.19	5.59	5.18	4.87	4.05	3.41	2.81
30	13.29	8.77	7.05	6.12	5.53	5.12	4.82	4.00	3.36	2.76
40	12.61	8.25	6.59	5.70	5.13	4.73	4.44	3.64	3.01	2.41
60	11.97	7.77	6.17	5.31	4.76	4.37	4.09	3.32	2.69	2.08
120	11.38	7.32	5.78	4.95	4.42	4.04	3.77	3.02	2.40	1.77

Table C.7
Critical values of Chi Square.

df	0.75	0.90	0.95	0.97	0.99	0.995	0.999
1	1.323	2.706	3.841	5.024	6.635	7.879	10.828
2	2.773	4.605	5.991	7.378	9.210	10.597	13.816
3	4.108	6.251	7.815	9.348	11.345	12.838	16.266
4	5.385	7.779	9.488	11.143	13.277	14.860	18.467
5	6.626	9.236	11.070	12.833	15.086	16.750	20.515
6	7.841	10.645	12.592	14.449	16.812	18.548	22.458
7	9.037	12.017	14.067	16.013	18.475	20.278	24.322
8	10.219	13.362	15.507	17.535	20.090	21.955	26.124
9	11.389	14.684	16.919	19.023	21.666	23.589	27.877
10	12.549	15.987	18.307	20.483	23.209	25.188	29.588
11	13.701	17.275	19.675	21.920	24.725	26.757	31.264
12	14.845	18.549	21.026	23.337	26.217	28.300	32.909
13	15.984	19.812	22.362	24.736	27.688	29.819	34.528
14	17.117	21.064	23.685	26.119	29.141	31.319	36.123
15	18.245	22.307	24.996	27.488	30.578	32.801	37.697
16	19.369	23.542	26.296	28.845	32.000	34.267	39.252
17	20.489	24.769	27.587	30.191	33.409	35.718	40.790
18	21.605	25.989	28.869	31.526	34.805	37.156	42.312
19	22.718	27.204	30.144	32.852	36.191	38.582	43.820
20	23.828	28.412	31.410	34.170	37.566	39.997	45.315
21	24.935	29.615	32.671	35.479	38.932	41.401	46.797
22	26.039	30.813	33.924	36.781	40.289	42.796	48.268
23	27.141	32.007	35.172	38.076	41.638	44.181	49.728
24	28.241	33.196	36.415	39.364	42.980	45.559	51.179
25	29.339	34.382	37.652	40.646	44.314	46.928	52.620
26	30.435	35.563	38.885	41.923	45.642	48.290	54.052
27	31.528	36.741	40.113	43.195	46.963	49.645	55.476
28	32.620	37.916	41.337	44.461	48.278	50.993	56.892
29	33.711	39.087	42.557	45.722	49.588	52.336	58.301
30	34.800	40.256	43.773	46.979	50.892	53.672	59.703
40	45.616	51.805	55.758	59.342	63.691	66.766	73.402
50	56.334	63.167	67.505	71.420	76.154	79.490	86.661
60	66.981	74.397	79.082	83.298	88.379	91.952	99.607
70	77.577	85.527	90.531	95.023	100.425	104.215	112.317
80	88.130	96.578	101.879	106.629	112.329	116.321	124.839
90	98.650	107.565	113.145	118.136	124.116	128.299	137.208
100	109.141	118.498	124.342	129.561	135.807	140.169	149.449

D Design of Experiment Tables

D.1 FACTORIAL DESIGN

Full Factorial: 2 factors 2 levels

run	A	B
1	+	+
2	-	+
3	+	-
4	-	-

Full Factorial: 3 factors 2 levels

run	A	B	C
1	+	+	+
2	-	+	+
3	+	-	+
4	-	-	+
5	+	+	-
6	-	+	-
7	+	-	-
8	-	-	-

Full Factorial: 4 factors 2 levels

run	A	B	C	D
1	+	+	+	+
2	-	+	+	+
3	+	-	+	+
4	-	-	+	+
5	+	+	-	+
6	-	+	-	+
7	+	-	-	+
8	-	-	-	+
9	+	+	+	-
10	-	+	+	-
11	+	-	+	-
12	-	-	+	-
13	+	+	-	-
14	-	+	-	-
15	+	-	-	-
16	-	-	-	-

Full Factorial: 5 factors 2 levels

run	A	B	C	D	E
1	+	+	+	+	+
2	-	+	+	+	+
3	+	-	+	+	+
4	-	-	+	+	+
5	+	+	-	+	+
6	-	+	-	+	+
7	+	-	-	+	+
8	-	-	-	+	+
9	+	+	+	-	+
10	-	+	+	-	+
11	+	-	+	-	+
12	-	-	+	-	+
13	+	+	-	-	+
14	-	+	-	-	+
15	+	-	-	-	+
16	-	-	-	-	+
17	+	+	+	+	-
18	-	+	+	+	-
19	+	-	+	+	-
20	-	-	+	+	-
21	+	+	-	+	-
22	-	+	-	+	-
23	+	-	-	+	-
24	-	-	-	+	-
25	+	+	+	-	-
26	-	+	+	-	-
27	+	-	+	-	-
28	-	-	+	-	-
29	+	+	-	-	-
30	-	+	-	-	-
31	+	-	-	-	-
32	-	-	-	-	-

Fractional Factorial: 4 factors 2 levels 8 run

run	A	B	C	D
1	+	-	+	-
2	-	+	-	+
3	-	-	+	+
4	+	+	-	-
5	-	-	-	-
6	+	+	+	+
7	+	-	-	+
8	-	+	+	-

Fractional Factorial: 5 factors 2 levels 16 run

run	A	B	C	D	E
1	-	+	+	-	+
2	+	-	+	+	-
3	+	-	+	-	+
4	-	+	-	+	+
5	-	-	+	+	+
6	-	-	+	-	-
7	+	+	+	-	-
8	+	-	-	-	-
9	-	+	+	+	-
10	+	-	-	+	+
11	+	+	+	+	+
12	+	+	-	+	-
13	-	-	-	+	-
14	-	+	-	-	-
15	-	-	-	-	+
16	+	+	-	-	+

Fractional Factorial: 6 factors 2 levels 32 run

run	A	B	C	D	E	F
1	+	+	+	+	+	+
2	-	-	+	-	-	+
3	-	+	-	+	-	-
4	-	-	-	+	+	-
5	+	+	-	-	+	+
6	-	-	+	+	-	-
7	+	+	-	-	-	-
8	-	-	+	+	+	+
9	+	+	-	+	+	-
10	-	+	+	-	+	+
11	-	-	-	-	+	+
12	-	-	+	-	+	-
13	-	+	-	-	-	+
14	-	+	+	+	+	-
15	+	-	-	-	+	-
16	+	+	+	+	-	-
17	+	-	-	-	-	+
18	+	-	+	+	+	-
19	+	+	-	+	-	+
20	-	+	+	+	-	+
21	-	+	-	+	+	+
22	+	-	-	+	+	+
23	+	-	+	+	-	+
24	-	-	-	-	-	-
25	-	+	+	-	-	-
26	+	-	-	+	-	-
27	+	-	+	-	-	-
28	+	+	+	-	-	+
29	+	-	+	-	+	+
30	+	+	+	-	+	-
31	-	-	-	+	-	+
32	-	+	-	-	+	-

D.2 PLACKET BURMAN

Placket Burman: 12 run

run	A	B	C	D	E	F	G	H	J	K	L
1	+	+	-	+	+	+	-	-	-	+	-
2	+	+	-	-	-	+	-	+	+	-	+
3	-	+	-	+	+	-	+	+	+	-	-
4	-	+	+	-	+	+	+	-	-	-	+
5	-	-	+	-	+	+	-	+	+	+	-
6	-	-	-	+	-	+	+	-	+	+	+
7	-	+	+	+	-	-	-	+	-	+	+
8	-	-	-	-	-	-	-	-	-	-	-
9	+	-	-	-	+	-	+	+	-	+	+
10	+	+	+	-	-	-	+	-	+	+	-
11	+	-	+	+	+	-	-	-	+	-	+
12	+	-	+	+	-	+	+	+	-	-	-

Placket Burman: 16 run

remark: Screening 15 factors in 16 runs involves perfect aliasing of pairwise interactions of the first six factors with the last factor.

run	A	B	C	D	E	F	G	H	J	K	L	M	N	O	P
1	-	-	-	-	-	-	+	+	+	+	+	+	-	-	+
2	-	+	+	-	-	+	+	-	-	+	+	-	-	+	-
3	+	-	+	+	-	-	-	+	+	-	+	-	-	+	-
4	-	+	+	+	-	-	+	-	+	-	-	+	+	-	-
5	+	+	-	+	-	+	+	+	-	-	-	-	-	-	+
6	-	-	-	+	-	+	-	-	+	+	-	-	+	+	+
7	+	+	+	-	+	-	-	-	+	+	-	-	-	-	+
8	+	+	-	-	-	-	-	-	-	-	+	+	+	+	+
9	-	+	-	-	+	+	-	+	+	-	-	+	-	+	-
10	+	-	-	+	+	-	+	-	-	+	-	+	-	+	-
11	+	-	+	-	-	+	-	+	-	+	-	+	+	-	-
12	-	+	-	+	+	-	-	+	-	+	+	-	+	-	-
13	+	+	+	+	+	+	+	+	+	+	+	+	+	+	+
14	+	-	-	-	+	+	+	-	+	-	+	-	+	-	-
15	-	-	+	+	+	+	-	-	-	-	+	+	-	-	+
16	-	-	+	-	+	-	+	+	-	-	-	-	+	+	+

Placket Burman: 20 run

run	A	B	C	D	E	F	G	H	J	K	L	M	N	O	P	Q	R	S	T
1	+	+	-	+	-	+	-	-	-	-	+	+	-	+	+	-	-	+	+
2	+	+	-	+	+	-	-	+	+	+	+	-	+	-	+	-	-	-	-
3	+	+	+	+	-	+	-	+	-	-	-	-	+	+	-	+	+	-	-
4	-	+	+	+	+	-	+	-	+	-	-	-	-	+	+	-	+	+	-
5	-	+	-	+	-	-	-	-	+	+	-	+	+	-	-	+	+	+	+
6	-	-	-	-	+	+	-	+	+	-	-	+	+	+	+	-	+	-	+
7	-	+	+	-	-	+	+	+	+	-	+	-	+	-	-	-	-	+	+
8	+	+	+	-	+	-	+	-	-	-	-	+	+	-	+	+	-	-	+
9	+	-	+	-	+	-	-	-	-	+	+	-	+	+	-	-	+	+	+
10	+	-	-	+	+	+	+	-	+	-	+	-	-	-	-	+	+	-	+
11	-	+	-	-	-	-	+	+	-	+	+	-	-	+	+	+	+	-	+
12	+	-	+	+	-	-	+	+	+	+	-	+	-	+	-	-	-	-	+
13	-	-	-	-	-	-	-	-	-	-	-	-	-	-	-	-	-	-	-
14	+	+	-	-	+	+	+	+	-	+	-	+	-	-	-	-	+	+	-
15	-	-	+	+	+	+	-	+	-	+	-	-	-	-	+	+	-	+	+
16	-	-	+	+	-	+	+	-	-	+	+	+	+	-	+	-	+	-	-
17	+	-	-	-	-	+	+	-	+	+	-	-	+	+	+	+	-	+	-
18	-	+	+	-	+	+	-	-	+	+	+	+	-	+	-	+	-	-	-
19	-	-	-	+	+	-	+	+	-	-	+	+	+	+	-	+	-	+	-
20	+	-	+	-	-	-	-	+	+	-	+	+	-	-	+	+	+	+	-

Index

U

V

W

Z

Printed in the United States
by Baker & Taylor Publisher Services